The New Digital Enterprise:

The Power of Low-Code

And Collaborative Design

To Transform Your Organization

∼∼∼

Ronald S. Kagan

Library of Congress Cataloging-in-Publication Data
Kagan, Ron
The New Digital Enterprise: The Power of Low-Code And Collaborative Design To Transform Your Organization, by Ronald S. Kagan.
     pages cm

ISBN 978-1-7378916-0-4

1. Digital transformation. 2. Low-code technology. 3. Business development. 4. Human-centered design.

Birdrock Press
P.O. Box 12153
La Jolla, California 92039
www.birdrockpress.net

Founded in 2004, Birdrock Press rides a different wave with writers whose passion for their subject matter that sets them apart. Birdrock Press offers special editions for academic curriculums.

Printed in the United States of America
First printing October 2021
R 4.06

for
Helen and Noah

Contents

Preface

Gathered around a conference table speakerphone, we listened as the senior vice president in charge of the International Operations division described the bank's dilemma: Foreign banks were gaining traction in the U.S., and it was critical to roll out a digital presence that embraced a vision of global reach. In response, a new software system had been designed and delivered to branches worldwide. It would capture all aspects of a branch's international business in a single, secure application. I inherited the system, written by an enthusiastic group of coders, as my first major project management assignment when I took over the internal IT group. The goal was to give managers the tools to streamline their decision-making with the ability to adjust products, fees, and rates to remain competitive.

The Tokyo branch was one of the first to receive the new system, but had yet to complete even the most basic budget screens. As headquarters awaited a progress report from the Tokyo branch manager, the head of International Operations, Bill, had more on his plate to deal with. It wasn't just the lack of numbers that confounded him, but the continued pressure from his higher-ups to keep expenses under control even as plans were being laid to expand the department's mandate.

"Just send me what you have and let's go from there," Bill said congenially to the branch manager, a Chicago native on a mandatory two-year rotation overseas.

"OK, another week, and it's on its way," the Tokyo branch manager responded, clearing his throat. "We may have some more questions for the programmer. But one week tops."

"Great," Bill responded. "Looking forward to it. I'll get IT in the loop. Thank your staff for me."

When he hung up from his call, Bill looked around the room. There were a lot of skeptical faces. The branch manager had been assigned to the Tokyo office just a year ago, and the branch was historically slow moving, with every decision seemingly needing the entire staff's involvement and input. It was slow and painful, and the new manager couldn't seem to change it.

"He definitely needs a hand," Bill said to us. "He's got a lot on his plate. What I'm hearing is we need a way to get their process going. So, let's give them our expert. I think I'll send our project lead over there. "

All eyes turned to me.

"I guess that's you," Bill said with a smile. But the problem couldn't be the program, I thought to myself. It was so simple; anyone could enter the numbers.

I was the new kid on the block; the previous project manager had flown the coop not too long before. When you inherit code, you're more or less stuck with its organization and approach—it was more of a maintenance chore than a system I could sink my teeth into. The New York branch wanted some bells and whistles squeezed in which weren't technically complex but created a mishmash of the program.

On the way out of the office, one of Bill's staff pulled me aside.

"Let's get some coffee downstairs," she said. I was still a bit out of it, thinking about what to pack and what could be going wrong in Tokyo.

"It's Yamata-san that you want to see," she said after we sat down. "He's the senior accountant and has been there for ages. On the org chart, he has a dotted line staff position. The branch manager rotates every two years and doesn't get to know the staff. Yamata-san is a fixture."

"OK," I said. "I can do that."

"But, when you see him, don't start out talking business. That doesn't work over there. Instead, talk about weather or your flight over or anything relatively lightweight."

"But Bill wants the numbers ASAP," I said.

"Yes, but the staff is stuck. And I bet that Yamata-san might have the answer."

Within days I was on a plane to Tokyo. I'd brought extra copies of the manual for the staff and had an appointment to meet the branch manager first thing.

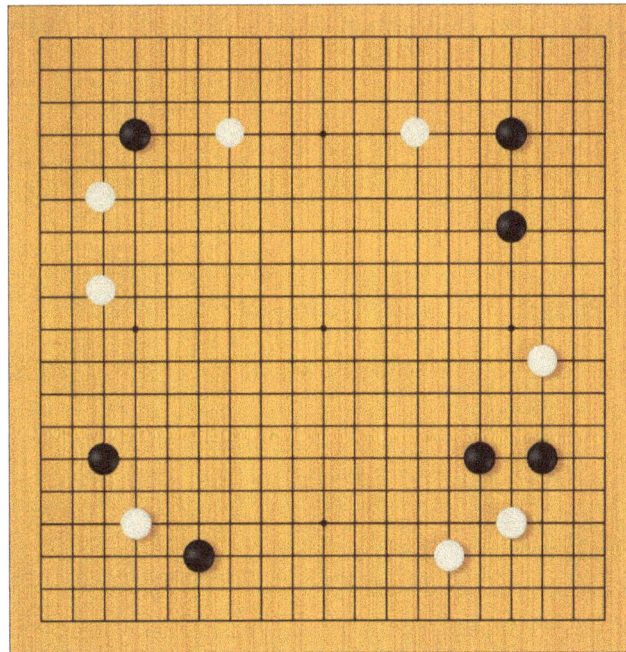

After meeting with the branch manager, I visited Yamata-san who was expecting me. He sat in a smaller office with a small window looking out to the busy main street. We talked about the weather and how the autumn leaves were just beginning to change. Then, for variety, I mentioned the Japanese game of Go. I'd learned to play in middle school and enjoyed it ever since. Yamata-san brightened up immediately—he too was a fan. He followed the national tournaments. I mentioned that what I really enjoyed is how go defied the typical way Western games were played. In chess, to capture your opponent's piece, you moved yours to the same square as your opponent's. In Go, you captured by surrounding your opponent's piece. Tea was brought in and we talked about joseki, opening moves in Go.

Near the end of my time with him, Yamata-san casually introduced the topic of the new system, and acknowledged the staff was having issues with it. I asked him if he know what exactly they were having problems with.

"It is better to speak with them directly," he said.

I nodded.

"Instead of asking what problems they are having, Kagan-san, ask for suggestions for new features. They are polite. They do not want you to lose face."

"But I would not be insulted," I said.

"They would feel bad about it. Ask what they like about the system," Yamata-san suggested. "This will be the best path."

So I agreed to approach it as a new version, a design session, where they could freely describe their suggestions.

After a brief introduction by Yamata-san, I stood before the 20-or-so staff members, pulled from different sections of the branch. The managers were not present for the most part, just the workers. Instead of honing in on why they couldn't submit their numbers, I decided to change the subject of my visit.

"We are now putting together a new version. And I'd like to hear what you like about the current system. What would you like to see in the new version? Your suggestions are important to us. That's why I'm here."

At first they were a little shy, but slowly one hand raised, then another. Yamata-san interrupted.

"He is American and this is how he likes to work. Please do not be bashful."

Suddenly, there was a flood of things that the staff suggested. The most obvious were simple changes like the format of the dates (from month-day-year to day-month-year) and different categories of input more in line with how they ran their operation. Most importantly, they asked for a special box next to the amounts they were less sure about. It was an unusual request, since the budgeting process assumed there was some variation. But the Tokyo staff took pride in their precision. With a checkbox to indicate approximations, they were free to enter and revise some of the lines of the budget later. When these boxes were clear, the budget was ready.

After the meeting, there were enough suggestions to get started on updates. I made a call to my staff back home, and they got to work on the changes. Within a few days I presented a prototype of the system. There were a lot of vigorous nods. The data needed to be put into an interim state (in this case, yen) before a final reformatting into a format that the head office could use. Within a week, the Tokyo numbers were completed by the staff. The branch manager was pleased, as was Bill, who emailed "congrats!" from headquarters.

On the flight home I realized that designing and coding a solution that actually worked was more than simply the technical design and requirements documents. It was all about working side-by-side with the end users, showing your prototypes as early as possible, and developing the solution in a collaborative way with the people who understood the work the best. My project team needed to get out of their offices and in front of people.

In many ways, my trip to Tokyo was a turning point in my career. A few years later I started my own software company and introduced products built with the involvement of the actual end users.

1. Introduction

The world of software development has come under pressure in recent years as companies engage in the process of automating their workflow and expanding their digital presence.

Traditional platforms for developing systems using low-level languages, full-stack coding, and relational databases have become cumbersome for most organizations. Companies can no longer afford the engineering talent, the time, or the expense of building everything from scratch. Additionally, work has become increasingly complex and specialized, and Information Technology (IT) departments do not have the depth of knowledge to adequately design the solutions needed by various departments. Understanding these details is critical to every facet of the design and development of software solutions: it determines whether the system actually works, or is a placeholder until a better solution can be found. Perhaps most pressing is the fact that classic approaches to software development using written specifications are too clumsy, time-consuming, and expensive.

In short, software development has changed, because the platforms, processes, and people have changed.

Sticking with the old ways may be adequate for now. But to keep up with a rapidly changing business environment, new demographics, and an increasingly dispersed organization, managers must embrace the newer way of doing things. To this end, there are three basic areas that give companies the advantage in the development of their digital solutions:

1. **Low-code/no-code software (LCNC)**. Instead of building from the ground up, new platforms allow software to be developed using, for example, drag-and-drop selections. Think of this software as akin to the spreadsheets that are already used in most organizations. *Low-code/no-code* platforms expand this approach with the suite of built-in, easy-to-use functions. Little programming is needed, and it promotes end-user development alongside technical experts.

2. **Collaborative design**. Instead of a select few designing a needed piece of software, the *collaborative design* process expands these groups and provides the steps needed to understand problems, provide innovative solutions, and roll out systems using prototypes.

3. **Digital design hubs**. If an organization is spread across multiple offices in different locations and a growing majority of employees work from home, the informal side of the organization can be mobilized to bring knowledge and innovation to the development process. As these networks become more virtually engaged, they form *digital hubs*, able to span both the informal and formal organization.

By combining these recent advances into a hybrid system of design, organizations have a more reliable way to develop their software while at the same time reducing time and resources needed on their projects.

This is the foundation of *low-code design hubs* and the focus of this book.

Low-code/no-code software

Low-code/no-code software (LCNC) is a new generation of software development platforms that require very little programming, if any. Traditionally software engineers develop software using programming languages that only seasoned coders know, and most programmers have their favorites, like C++, Javascript, Python, Java, or C#. It takes time and effort to build software, usually involving constructing code foundations, adding layers and libraries, then building out the programming objects and their interfaces, and mapping their instances to fields.

Drag and Drop Components

With LCNC, instead of building a piece of software from the ground up, LCNC has a complete set of objects and functions ready to go, and utilizes simple instructions. Since these instructions are easy to understand, the barriers for entry are much lower. Now everyone and their mother can program—kind of. In reality, it's actually giving instructions to the LCNC platform that translates it into a language the computer or mobile device understands. The same instructions get translated into different screen formats so the organization can use it on mobile devices, laptops, or desktops regardless of type. It can be used on your PC or iBook, iPhone, or Android. Imagine all the doors that this new way of programming opens up. And many times you don't even need to give instructions: you can design a screen or report by dragging the field over to a screen and positioning it where you want it. The LCNC platform handles the rest.

LCNC simplifies the development process, allowing individuals with minimal coding training or experience to write software. If it's not obvious from this description, LCNC can revolutionize how an organization handles its systems work. Costs and development time are dramatically reduced. The software is more apt to work correctly since the staff is actively involved in its design and coding. IT departments now have a streamlined way to develop solutions for other internal departments. Companies can use this for their clients and customers, providing ways to interact with their mobile devices or computers that they did not have before.

The LCNC platform also addresses staffing issues in IT departments. There is a shortage of qualified programmers and analysts, with increasing workloads as more organizations undertake digital transformations. That is, companies now move a good deal of the work into software, and allow their staff and customers to use this instead of direct staff contact. In the long run, LCNC saves the expense of employing and training a larger staff.

LCNC requires no formal software education and no engineering background, as libraries and foundations are pre-built. Companies are now turning to this mode of development, and software vendors are building LCNC capabilities into their core products. According to Forrester Research, LCNC software is expected to reach $21.2 billion in global sales by 2022, up seven-fold from $3.1 billion in 2017.[1] This growth will continue, reaching a projected $30.4 billion — a ten-fold increase — by 2026.[2]

Collaborative design

LCNC holds great promise for all sizes of organizations across a variety of fields, from the financial industry to retail. As with any software, the key to its success is in its design. When LCNC is set loose in an organization, it can still produce buggy, inconsistent solutions that don't work right, or fail to address the important issues the company is trying to solve. The rollout may be quick, but the application may be sick.

Identify

Refine   Brainstorm   Select

Plan   Build   Test

Close attention should be paid to how the software is designed, tested, and implemented. Gone are the days that you can code up an application and put it into production. Security, privacy, and access must be considered, as well as potential exceptional circumstances. Thankfully, recent innovations in collaborative design processes hold the answers to how the design process can be expedited with the right people taking the right steps. The software is no longer solely in the hands of the programmer, but rather now involves staff who can directly code portions of the final product. And it's no longer a special team appointed by management to be responsible for the software, but the involvement of informal groups, or *hubs*, who worked closely with one another and know the work and customers the best.

Combining software development with collaborative design steps, along with the involvement of these informal hubs, leads to a better deployment of software. These are the ingredients of software development moving forward. The seven steps of the design process can be summarized by the diagram on the previous page.

Though COVID has greatly impacted our ability to come together as a team to design, engaging staff across departments in virtual meetings provides the responsive environment that organizations desperately need. Part of the problem all organizations face is uncertainty. There is uncertainty in the marketplace, with staff, with shifting values and resources, and with our ability to respond accordingly. The informal networks accommodate these changes.

When viewed from an organizational behavior standpoint, these lateral processes across departments and formal project teams provide an efficient way to handle uncertainty. Studies on the effectiveness of these informal processes and their collaborative nature have been the subject of behavioral research for a number of years.[3]

LCNC will not necessarily solve all your problems, but it possesses the ability to ramp up solutions. The same care must be taken to analyze the issues that the software is aimed at solving (for example, handling increase volumes of customer transactions), just as you would with traditional solutions. Many times, the solution can be simply buying a software package (see *Why Not Just Buy Some Software?* on page 95) "The obvious question is why not purchase off-the-shelf software or subscribe to a software service (SaaS) to address the issues? In many ways this can be an excellent solution. The cost is known upfront, and a support staff is available from the company to handle any questions or bugs." . A specialized package might solve 60% of the issues, and this may be enough. Or you may find that leaving things as they are is acceptable. It is important that a design team look into possible solutions. The collaborative process itself will strengthen the team and allow for a more detailed review of the issues at hand.

Digital design hubs

Organizations have a natural ability to optimize their performance and enhance their efficiencies. In his research into social interactions, MIT professor Alex Pentland demonstrated the natural flow of ideas and engagement that transcend the formal hierarchy.[4] These collaborative networks get the work done: they are able to fine-tune workflow as the needs and goals of the enterprise change; they have the knowledge of internal resources and the ability to facilitate a freer flow of information between departments and with clients; and they can empower collaborative centers.

With the introduction of new LCNC technologies in the workplace, team members can often fashion their own reporting, screens, and workflow with minimal knowledge of software programming. They can also do this remotely with the use of virtual meetings, e-mail, and phone calls. Members of the IT team can be brought in to bridge the inherent complexities that might arise, as well as ensure that security and access follows the prescribed procedures.

These informal centers form the primary groups in the design process as well. As the aforementioned *hubs* are collaborative by their nature, but have the guidance and support of upper management, they are formed to drive the organization more effectively. Whereas a formal team may need to obtain permission from the managers of other departments to undertake a particular task, hubs can reach out across departmental boundaries and get the task taken care of.

When these hubs are digitally connected using online communications platforms (as described in more detail in *Online communications* on page 90), they become even more effective, as they can engage portions of the organization that are in different locations or that even work remotely. The technology used can be something as simple as texting between members, or being linked by forums or social media. For the enterprise, unique technologies are evolving to keep communications private for the hub while promoting engagement across departments and business silos. Hubs can also span organizational boundaries forming support groups, wherein both customers and specialists interact to answer questions and provide customer support. With these technologies in place, these groups become digital hubs. When they are brought into the development process, they are referred to as *digital design hubs*, and include other teams that help guide the effort and inform the larger organization. (See *Digital Hubs* on page 53 for more details).

Engaging multiple centers of an organization through digital design hubs increases the efficiency of the whole enterprise, including the reduction of costs and long-term maintenance. When conflicts arise, facilitation can be used to overcome the impasses. Yet the challenge still remains in choosing and designing the correct solution for the organization, as well as assembling and running the design teams collaboratively. This book explains the design process used to build LCNC solutions using digital design hubs and collaborative design frameworks.

2. How to Read This Book

This book is meant to be a practical guide to implementing low-code/no-code solutions using collaborative design techniques. It is divided into sections, and the reader is encouraged to jump around based on their interests.

Chapters 3 - 5 cover the changes happening all around us. These are both external, e.g., our business environments, and internal, e.g., our workforce and the company environment.

Chapters 6 - 8 discuss the formal and informal structures inside an organization, and how hubs are at the heart of how a business functions. Its members may have different approaches to the formal procedures, but they all know how the company works. For that reason alone, they are critical to the automation of these processes.

Chapters 9 - 10 outline the collaborative design process by which groups more effectively design their systems using collaborative steps. These steps are examined in detail and focus on the people involved in the work.

Chapters 11 - 12 introduce low-code/no-code (LCNC) platforms. They offer new ways to design and code systems that streamlines implementation. When combined with collaborative design and hubs they can be used to prototypes solutions.

Chapter 13 - 14 details how digital design hubs are composed and rolled out into the collaborative design process to develop solutions using low-code/no-code platforms.

Chapters 15 - 20 walk through the process of designing and implementing a solution. The steps analyze the basic documents and reports that are needed, workflow, approval processes, and customer systems. Since not all low-code/no-code platforms are alike, the chapters also cover selecting the right platform for your organization.

Chapters 21 - 23 focus on rolling the final system into production and the approach to maintenance.

Chapter 24 summarizes the major topics in a Quick Guide format for designing and developing low-code/no-code software using collaborative design.

Thinking outside the box is encouraged when considering new solutions for an organization. This book walks the reader through a design and implementation process using collaborative design techniques with hubs and low-code/no-code platforms.

Part I: The Landscape

3. The Age of Moving Targets

If you've ever tried to deliver on a moving target, that is, deliver a design or final product when the design itself keeps changing, then you're probably not surprised that almost everything these days is a moving target. Just when you think you know what your customer wants, they pivot. Maybe it's a little to the left or a little to the right, or maybe it's second thoughts a week before delivery of a custom solution. More often than not, the early phase of projects swing wildly and widely in their goals, their technology, their aesthetics, and of course the pricing and personnel that make it all happen.

Dealing with these moving targets is both a skill cultivated from experience and the ability to recognize a project's pivot points. Pivot points are fixed: everything else may revolve around them, but these points are the common ground.

There are a number of factors that contribute to the motion of organizational goals, product designs, and decisions both big and small. But underneath most of these is a new level of uncertainty. Whether it begins in the business environment or with changes in staff or procedures, uncertainty seems to spread from one department to another, from one product line to another. The ability to address it becomes more complicated as organizations and their technology are less able to handle the new demands, or react in a clear way to the ambiguities that arise. Historically products evolved over a number of years; now change is the norm, and we must quickly adapt to these newer technologies.

> "All the layers and specialization are breaking down. Instead of a year, we want to put an idea in front of a customer in a week."
>
> -Douglas Safford, Allstate's vice president of technology innovation.[5]

Enterprise is a balancing act of demands from customers, leapfrogging of competitors, disruptive technologies, and global changes in demands, regulations, and tastes. The work itself has become more complex, requiring multiple disciplines and specialties, flexible timelines, and cohesive teams. And it is increasingly performed by a workforce whose culture, demographics, and incentives are shifting.

For an enterprise to hit a homerun, it must first find the plate. It must be able to focus on the wind up, and it must be agile enough to react quickly as an opportunity comes flying down the lane.

Globalization

Products and services compete globally. Even if they stay domestic, there is increased pressure from overseas competitors. Goods must adapt to a diverse customer base, while also dealing with shortened life cycles of the products themselves. Take DuPont as an example: two decades ago their products had a 20-year lifespan; now it's a mere two years. Businesses must keep the current customer base while expanding to new ones, and also understand these local markets and their requirements.

In foreign markets, products must meet the local requirements, prioritize features, and respect the marketing approach of that country. This increases the variations of products and services. Not only must languages be integrated into the final products, but localization of currencies and other cultural aspects must be considered.

Unsurprisingly, the staff who are most likely to understand local markets are located in the countries themselves. To reduce costs they may be integrated as partners, but accommodating these resources and personnel requires active planning and coordination. As markets evolve, the appropriate adjustments must be made.

Complexity

Imagine producing an animated feature film. This process contains many overlapping moving parts: an understanding of the intended audience; the involvement of writers, producers, artists, technicians, colorists, quality assurance teams, and technical workers; and the support staff needed to integrate all of these distinct teams seamlessly into a productive workflow.

The final product is not simply writing the narrative, developing the characters, and producing the animation. It is an iterative process where narration, characters, and even technology evolve over the course of many years. Technology products also have an iterative process, and rely on a variety of players in every stage of their life cycle.

To handle the increased complexity of the work and the sometimes fluid set of requirements, teams often must be composed of specialists. Their *complexity* is reflected in the number of independent individuals interacting with one another on a number of levels.[6]

This can be done with the work teams themselves being defined within the organization to address the design, development, and support of the final product.

Conversely, the organization may create a procedure wherein specialists are brought on board at various stages of development and production. In either case, the ability for a workforce to interact with other fields is important.

For example, if a software product relies on a database housed in cloud storage, the software engineers may need to bring in a database administrator or designer to assist with the final product. The needs of the database administrator may be quite different than those of the software engineers. Although both are part of software development, their focus, terminology, and workflow made be entirely different.

The underlying complexity of projects dictates how teams interact and organize themselves. Products and services themselves are complex as they require involvement from different players.

Customer relationships

Work is increasingly dependent on establishing and maintaining customer relationships. This is especially true of large customer bases (e.g., governmental agencies, hotel chains, multinational companies) but also important for smaller entities (e.g., local businesses, family restaurants). At the core of this dynamic is a customer's reliance on the recommendations and work of an organization. This can be effectively provided when a basis of trust and personal rapport is established.

As staff in organizations and their client base change, so too do the customer relationships. An organization must be able to adapt quickly to these changes as well as maintain long-term relationships even as its own staffing evolves.

Technology

Where processes and products incorporate or use different technologies, these technologies adapt to meet evolving demands. For example, streaming services were historically restricted by the bandwidth on mobile devices and the cost of Internet connections. As WiFi becomes more readily available at higher bandwidths and accessible price points, the delivery of hi-fidelity audio and high-resolution video has changed. This has impacted the various formats used to deliver audio and video, as well as the storage requirements.

Technology incorporated into other products or services also go through similar changes. Operating systems themselves evolve and require applications to upgrade. Processor and graphics capabilities continue to push the envelope, and with it their impact on the products and services that rely on their underlying technology.

Gig economy

The trend across multiple sectors of the economy has been for the work to be performed by subcontractors, either individual contractors or businesses that carve out portions of the final product or services to focus on. The advantages are a lower cost in general, as the outsourcing avoids employee benefits, the extra overhead of HR, and a more responsive workforce based on the task load. It also presents disadvantages in consistency across a workforce, less cohesive teams because of the revolving door, and decreased engagement of the subcontracted personnel.

Customization

Customers depend on companies to fine-tune their products and services for their special requirements. These customizations or enhancements are critical to the customer, so the work itself must accommodate the economies of scale for the base product or service, but adapt quickly as specific customizations are needed.

This complicates the work, since those areas that may need to be changed in the final product must be known in advance. Or, at a minimum, the product or service should have enough flexibility built in to permit either the final user or a specialist at the company to change it.

Final thoughts

There is much to consider in today's work environment, both in things in our internal environment we may be able to predict and those that are outside, external forces beyond our immediate control but we must be resilient enough to adapt to. Changing demands from customer and consumer bases, evolving technologies that force updates to existing ones, and transient work forces with iterative teams and teamwork are redefining what it means to work today, and to effectively roll out a satisfactory—even an excellent—product or service.

4. Where's my Desk?

The workplace has changed dramatically over the last decade. What used to be a single office setting with a traditional 9-to-5 schedule has now spread across multiple locations and time zones. It has blossomed into a rich, multi-cultural, multi-generational amalgam. Telecommuting became a requirement during the pandemic, but also became a means to enhance the work environment and productivity.

Yet the ability to sustain and engage both in and with this work environment comes under the constant pressure of a changing marketplace, more complex products and services, global markets and extended time zones.

Finding a way for people to sustain their focus and be engaged is the challenge of modern organizations. More importantly, what used to be conversations around the water cooler have now moved to e-mails, direct messages, and tweets. The ability of teams to know one another is diminished by the simple nature of not spending time socially with one's teammates. Connecting now is making any sort of contact with teammates outside of the formal e-mail and discussion forums; it's difficult to find common ground when you know little about who sits outside your department. Technology has become an expedient way to try to build trust between members.

Demographics

Over the last 20 years, there has been a significant shift in the structural composition of the workforce. The shift is due in part to the urbanization and migration of a younger workforce. Millions of people have migrated from countries with fewer opportunities to more prosperous regions. The U.S. has an estimated 13 million permanent immigrants as of 2011, with the Department of Labor expecting immigration to reach 80 million by 2050.[7] This influx brings a younger demographic as well as cultural diversity to the workforce.

In addition, it is estimated that over 22 million women will enter the U.S. workforce by 2050, with tertiary education levels exceeding those of their male counterparts, meaning a significant proportion of high-skilled positions will be held by women.[8]

According to the U.S. Bureau of Labor Statistics, population growth was at a projected 0.7% annual rate from 2010 to 2020 as compared to 0.8% for 2000 to 2010 and 1.3% for 1990 to 2000. IDC expects that 39 % of the workforce will consist of millennials with "instant communications, video on all devices, and seamless connectivity".[9] With the 55+ group increasing within the labor market, U.S. Bureau of Labor Statistics projects the prime-age workforce (ages 25 to 54) will drop to 87.3% of the workforce by 2029.[10]

Organizations will look to the younger Millennial generation (born between 1981 and 1996) to replace these jobs. This generation is even larger than the Baby Boomer generation, yet this group may not be ready to replace retiring workers at these higher-level roles that require specialization and education. Instead smaller Generation X (born between 1965 and 1980) may step into these roles.

A 2012 survey by the Pew Research Center found that the majority of this younger generation (those born between 1981 and 1996) feel they do not have the necessary education or training to enter the workforce. This presents a further obstacle for filling the positions vacated by a retiring generation.

In addition, there is increased ethnic diversity in the workforce, reflecting the changing populations in the U.S. This adds a different component to the ways teams and groups interact at work. Since cultural background plays significant role in values and norms of behavior, teams composed of multiple generations and ethnic backgrounds must be sensitive to differences in etiquette, values, and social norms. And yet, there is a clear advantage to this shift: given global markets and the international offices of many companies, the diversity of today's workforce provides an increased understanding of the needs of these markets and international groups.

Mobility

Traditionally, sales forces have been the main areas of a company that have been mobile, servicing the regions that they have been assigned. But with telecommuting, dispersed offices, and extended work hours, the mobile workforce within companies has grown dramatically. According to forecasts from the International Data Corporation (IDC), U.S. mobile workers will reach 78.5 million by 2024, representing 60% of the total workforce.[11]

This means that more and more employees rely on their mobile phones and devices throughout the workday. In a survey conducted in 2016 by the Gartner Group, 38% of companies expected to stop providing devices to employees, leaving them to utilize their own personal devices for work matters.

Global

According to the Bureau of Economic Analysis (U.S. Department of Commerce), worldwide employment by domestic multinational companies increased 1.5% in 2011 to 34.5 million workers, primarily abroad.

In addition to domestic branches, companies continue to spread out. The advantage is that separate regions can be more effectively serviced, and customer relationships enhanced by a geographically closer group.

Work-life balance

As the number of couples in the workforce where both individuals work outside the home increases, there is a growing need to accommodate the needs of couples to provide care to their families. National Alliance for Caregiving estimates that 65.7 million people serve as unpaid family caregivers to an adult or a child. According to a 2010 study conducted by MetLife, this ends up costing businesses $13 billion annually.

The impact on the workplace is two-fold: some employers now provide on-site daycare and flextime to accommodate the needs of parents, and they also provide family leave, where partners are paid while they care for their newborn or family member.

According to the National Study of Employers (2012), flex time has increased from 66% of companies surveyed in 2005 to 77% in 2012, with 63% working at home for some or a majority of their work time.

Final thoughts

Workplace dynamics are rapidly changing, encompassing multiple generations and cultural backgrounds. A recent survey by Society for Human Resource Management (SHRM) showed that 35% of respondents indicated that their companies are training line managers to respond to generational differences, and 32% are implementing diversity education programs.

5. It's Somewhere On This Screen

Information lies at the heart of any efficient organization. Getting relevant and up-to-date information in the right format and in a timely fashion is critical to making the right decision. It is also fundamental to efficiently implementing needed solutions.

Organizations have evolved to incorporate a multitude of systems to support a wide variety of applications. The finance department will have an accounts receivable and payable system. Human Resources will have an HCM system. Sales will have a CRM system. IT will have their IDE environments to develop their systems. Marketing will have social media and web platforms. Management will have its management reporting and budgeting systems. And of course to link the entire business together, an organization will implement a series of e-mail and scheduling systems, or even provide a proprietary social platform to keep everyone informed of other departments' happenings.

Did this have to be an e-mail?

When the types of decisions are straightforward, discussions naturally evolve and resolve in e-mail correspondence. The subject may start out in an outlying issue and turn into a more complicated description of the problem, or skip to implementing a solution. Key managers will be cc'd or directly included in the e-mail to solicit approval to move ahead. Once approval is given, teams move on to implementation, keeping the stakeholders involved along the way. However, e-mail has its own issues, inherent to its structure:

- It is part of the formal organization, so it serves as a record of positions, discussions, and decisions. So, it is hard to encourage the free flow of ideas when everything is on the record.

- It lacks the structure to analyze problems or brainstorm possible solutions. A thread may start on a symptom of an issue and never make any discernible progress toward identifying the underlying cause.

- Roles are still in place. That is, the project lead's e-mail on a topic will trump a coworker's, even if the coworker may have better insight. E-mail usually follows the hierarchy of the organization, making it difficult to gain a new perspectives on a given subject.

- Records of discussions may in fact contain important information or decisions. Yet, these must be recovered by going back through the threads. In doing so, new conclusions may be discovered.

- E-mail can of course get deleted, filed incorrectly, or caught in spam folders.

- With the proliferation of spam, inboxes get filled with junk. Even with the best spam filters, unwanted e-mail is mixed with important. Personal e-mails get mixed with business related ones. Subject lines become the way to sort by topics, but even these are inadequate to group related e-mails together.

- E-mail does little to provide concise analysis of a subject. The analysis can be spread across the thread.

- Recipients and senders will vary, yet there is no grouping of these into a single thread for review.

In short, e-mail forms a digital foundation of the formal organization. It insures that user departments and staff are involved in the conversations, and it also becomes a system of record. It can help jumpstart a topic and in many cases, it can provide discussion. But it is not conducive to a nimble organization, where change is ongoing. What may have been relevant a month ago needs closer attention now. To analyze this in a folder of e-mail becomes both time-consuming and inefficient. To replicate the organization's formal structure in the formal chain of e-mail hinders the possibility of a flexible team, where those closest to the problem have a stronger voice in defining issues and suggesting ways to proceed.

Report-itis

Many organizations take pride in the cornucopia of reports available throughout their organization. Yet even these take time to analyze. And when multiple departments must put their heads together, the information must be understandable to novice users.

Reports offer a tremendous source of information. But if they are in a set format, they may be harder to use initially, and may not provide good analysis in evolving situations. An interactive report, such as a sortable spreadsheet, offers an improved approach. But there are disadvantages to amassing reports for the new organization:

- The source of information remains the same even when new sources are present.

- Assumptions about the underlying numbers may detract from a realistic view of the over-arching information and situation.

- The ability to apply more involved filters, to sort, and to search criteria is limited.

Without direct access to the databases that feed the reports, basic reports can become obsolete quickly. Further, they can reinforce a perspective on a goal or issue based on the report's format and organization. For example, to try to determine why a group of customers ordered fewer products over a given quarter, a sales report will show the number of sales and the types of products. But regardless of the sorting mechanisms available in the report, it may not shed any light on the causes. To gain this sort of insight, management would need to drill down into the underlying databases to perform some sort of regression analysis on the quantitative characteristics of the customers and products.

Another approach is to bypass the reports entirely and rely on the insights on those closest to the customers: the sales team or the support team. This is a collaborative approach. Hubs focused on an issue like this still need data to fuel their analysis, and reviewing static reports may not be the best way to solve the problem.

Uncertainty and ambiguity

In situations of high complexity, organizations face an overload of information; combined with a high rate of change, it is not always clear what information is needed. In these cases, complexity of the work is connected to ambiguity rather than uncertainty. Unlike uncertainty, which is the *need for more information*, ambiguity of information implies that *several interpretations of the data and issues are possible*.

To solve problems in this arena requires a collaborative effort with team members closest to the information who are more adept at interpreting the information.[12]

Final thoughts

In the modern working environment, screens are unavoidable and there is a seemingly endless amount of information, input, and reports available at our disposal. Some of these forms of communication are more useful than others. The key is finding ways to self-sort and create the most (or more) efficient use of time and resources to avoid burnout, overload, and confusion.

Part II: Digital Design Hubs

6. Boundaries: Crossing Over

How an organization handles uncertainty, both within and outside the office, is a function of its structure. Most enterprises look strictly at their formal structure, that is, the divisions and departments laid out in their formal org chart, to ensure efficient workflows.

An individual is assigned a manager, who in turn also has a supervisor. Committees and projects have leaders, also assigned. And the organization as a whole sets up its incentives to gain control over how the work is executed, weeding out workers who do not fit the bill. As such, formal boundaries are a natural occurrence in any organization. They are vital to organizing work, focusing the right expertise on the tasks at hand, and maintaining an ongoing discipline to deliver quality and timely services or products. For example, the accounting department reports to the CFO, who in turns shapes the approach to finances the enterprise will utilize. These decisions are made by this department and its members contribute to the accuracy the company needs to meet its financial goals.

For work that is repetitive, this formal structure works well, with rewards in place for productive behavior. But by its very nature, this also cultivates an *us vs. them* mentality that can slow down operations and keep vital information from reaching those that are the most qualified to act on it. Turf wars can be waged to make sure other groups do not take on tasks assigned to a specific department. Especially at the higher echelons of a company, goals are set with higher production, lower costs, and year-over-year growth.

As the work becomes more uncertain or complex, handling situations requires cross-departmental teams that incorporate a broader range of specialties. Despite the interdepartmental nature of some specialized teams, if they are still divided along formal org chart lines this can still render them ineffective.

> Take for example, the IT department. Different sections of IT handle different sets of assignments, with the Customer Service group assigned to take calls from users with issues. These calls may be anything from computers failing to start, to bugs in programs written by other members of IT. Customer Support staff is rewarded based on the number of calls it takes to resolve problems, the rapport that they develop with callers, and how many calls they handle on average daily. Yet the resolution of the issues often falls outside their field of expertise.

> In these cases, the support representative may place the customer on hold and confer with other members of the section. When this fails however, the section must reach out to other members of the IT department. The issue can be logged and information taken, but the allocation of the work may fall on the shoulders of other sections with more intimate knowledge of the problem. The level of complexity of the work at hand, in this case solving a computer issue, directly affects the involvement of other specialties, many of which are located outside the formal section.

To this end, organizations need ways to allow boundaries to be easily crossed, even if it appears to go against the culture of the enterprise.

Formal vs. informal

Organizations have both formal and informal boundaries. The formal boundaries of the organization separate the enterprise from its suppliers, clients, and competition. The organization is further divided into departments and divisions along formal lines which serve to frame the expertise and level of authority that each has.

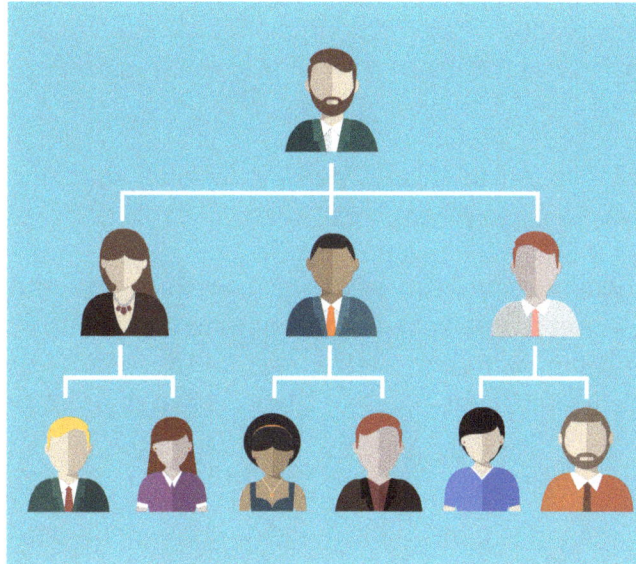

By contrast, an enterprise is also divided by less apparent informal lines. Employees that start with the company at the same time may get to know one another, creating an informal collective based on this shared experience. In many ways, these collectives form complex adaptive systems within the organization. For example, staff could work in parallel with common goals and tasks, but at varying levels of seniority. They adapt to the special circumstances to optimize the work with their own skill set and with future goals in mind. Their view of future events influence how they work with those around them. As the work becomes more complex and customized, these collectives can be characterized by *perpetual novelty*.[13]

In these situations, informal communications between members becomes the most effective and efficient way to understand and resolve problems.[14]

Other groups may form around common interests (e.g., working out at a local gym or taking classes together) or being at the same location. These may be a small, informal circle of acquaintances. Or, they may be larger networks that span departments and the organization itself.

Inside vs. outside

Organizations serve their employees, clients, and customers. For enterprises which must provide products or services to their client base, the operations are oriented to understand the needs of its clients, formulate a plan to service them, and then implement it to produce their final products and services.

More and more, companies are providing ongoing ways for their staff and customers to sit around the same table and interact. The advantages are:

- **More responsive environment.** As a customer's needs change, the closer ties allow the enterprise to readily adapt to the customer's requirements. Many times this takes immediate effect with new versions of products or changes to the manner of servicing the customer. Other times, the changing needs allow for the inclusion of more custom aspects of the enterprise's services.

- **Closer relationships.** The meshing of inside and outside groups can also allow for closer interactions between clients and their counterparts within the organization. For example, a sales team can get to know their clients on a more personal level, so the feedback and advice they may receive from their clients is clearer.

- **Better cultural understanding.** It is not just simply the products and services of an organization that benefit from these ties, but the enterprise understanding the cultural milieu of its client base.

- **Increased continuity.** As personnel change on both the customer's and company's teams, the increased contact allows other members on both sides to be able to adjust in a shorter period of time.

- **More efficient solutions.** As issues or obstacles arise, there are fewer barriers to addressing them head-on. With teams actively involved across organizational lines, they may be able to initiate solutions without having to involve higher levels in their respective organizations.

As the number of client groups that interact closely with their internal counterparts increases, the company as a whole is able to survey the market and have a clearer picture of what is needed. These groups may form around customers and be organized by region, product, or type of activity. Or they may involve suppliers and outside entities, e.g., technology companies such as Apple and Google, while others include app developers into the rollout of various software products.

The degree and frequency of the interactions between individuals on either side of the organizational boundaries can vary. As the frequency and influence of these groups increases, they may be formalized by the management of the companies.

Collaborative networks

When informal groups within an organization work together effectively, they compose *collaborative networks*. These networks are the building blocks of hubs (detailed in the next chapter) and can improve the efficiency and effectiveness of the organization. They also become critical to new innovation.

Yet attempts to harness these informal networks and to use them in the operations of a business have not always been successful. It is one thing to mandate a course of action or to set priorities through the formal organization. But when relying on an informal network, there are pitfalls.

In 2004, Boeing announced plans to build its Dreamliner 787. Instead of relying on its traditional approach of designing everything in-house, it set out to collaborate with its global suppliers, offloading the expense of both the design and testing onto its suppliers. Each supplier would be given a set component that they needed to produce and a timeframe. The hope was that this innovative approach would shorten the development time needed to produce the Dreamliner, and ultimately reduce overhead costs.

In a series of well-documented delays, planes were delivered four years late, and according to *The Wall Street Journal* Boeing lost close to $20 billion by the time its Dreamliners were profitable[15]. In the eyes of management experts brought in to rectify the situation, Boeing had simply failed to build an effective collaborative network with its global suppliers—each went its own way.

The ability for organization to collaborate across formal department lines becomes more important when there is a change of strategy or in the business environment. Yet there is a tension between the formal and informal boundaries of an organization. Formal boundaries dictate the procedures to be followed, the performance criteria for raises and promotions, and the assignments of responsibility and authority. In most cases, a member of one department does not have the authority to make decisions for another department. Steps that are prescribed by one department are not the ones followed by another.

Yet when work falls between the cracks—a situation that needs immediate attention—it is the collaborative networks that usually step in. Members of these networks know one another and can make time to help out. Management may further recognize the value of these collaborations and provide incentives and authority for them to act. However, these networks consume some level of resources, if only the time devoted outside their normal workload to help other sections.

Informal authority

When the work done by the collaborative networks is elevated by management, it can bridge the informal-formal divide. Rewarding the work of these teams is one way formal organizations incentivize collaborative networks. For example, management reviews may include managers outside their chain of command. If an accounting manager mentions an IT staff member who fixed an out-of-balance glitch in the G/L entries, that employee would get the attention of their supervisor. It would have a net position impact on the employee's future reviews, raises, and promotions. If an informal group of IT and operations staff weighed in on a customer issue, their members would receive higher ratings. Many times promotions into leadership positions look for collaborative talent from employees who reach out to and work well with other departments, and thus have a better understanding of the needs of their staff.

As collaborative networks become more central in resolving issues or laying the groundwork for future products and services, management may form its own informal team. The team can be composed of staff and managers who interact collaboratively around a goal. These teams (referred to as *advance teams*, discussed in the next chapter) include managers from various levels and departments with authority over their particular domain. For example, an IT supervisor may work alongside a sales manager to improve the sales cycle. Both hold a certain amount of authority in their areas. When they sit informally over lunch to discuss ideas, they carry with them an ability to authorize next steps. They may not be the exact ones to design and implement a solution, but they can be connected to the project. Advance teams provide a context for design hubs and can interject informal authority into the process.

In this way, informal networks cross over into formal reward structures, bringing an informal group into the formal realm. Their responsibilities do not formally include those tasks outside of their specialty, but their involvement with others outside their formal group is encouraged.

Levels

Informal groups may cross hierarchical levels in an organization. If an assignment goes beyond the authority of a certain team, it may get bumped to another team who can take action on it. Similarly, if a given project requires involvement of other departments, managers or supervisors may reach out to other team heads. Typically, an issue is relegated to the next level up in a formal organization with the supervisor deciding how best to tackle an issue; however, if a manager handled those types of issues in a previous position, they may be the best person to fulfill the request now. When time is critical, following the formal, rigid hierarchy can take excess time and complicate the response. Traversing levels has the advantage of respecting the mandates of the department (its turf) while expediting higher-priority decisions or answering more complicated questions needed in the work.

As levels become more rigid, the informal means of jumping over to another department's higher level may be discouraged. This is completely understandable when these requests are not too frequent. But if this becomes a common occurrence, managers should in fact encourage interaction from other levels in the organization. There is much to be said for the contact with staff at various levels.

- **Organizational culture**. When staff can speak to others at lower levels of the organization, they get a sense of the culture at that level. Issues that cause stress or conflict can be picked up on and discussed. This doesn't always happen on the first few conversations, but as rapport is established, more of the organization's culture can be understood.

- **Work-related issues**. When questions are asked about how to handle certain work situations, managers at a higher level should make sure that supervisors are brought into the discussion. Jumping over middle managers or project leaders can cause their own issues, so management must be prepared to listen and refer the matter down the line. But exposure to an issue gives managers insight into the types of things that slow down an organization or cause its workers stress.

- **Support**. When managers contact staff at other levels to express their support and show interest in the progress, the net effect is to enhance the flow of ideas, to incentivize staff, and to improve the overall culture.

- **Suggestions**. Managers can also reach out for verbal suggestions. Talking with workers and staff is a way to get input about a variety of topics: working conditions, issues, better ways of handling situations, and even general complaints. Again, it is the listening that counts. Where there are good suggestions or items that need attention, managers at higher levels in the organization can delegate a follow-up.

Final thoughts

In general, hierarchical levels in the organization serve a needed purpose to focus on the running of the business. That said, interacting with other levels when done in a transparent way has the advantage of giving managers a better understanding of issues and organizational culture.

As these collaborative networks take on increased importance in resolving issues or developing solutions, they segue into becoming the centers of the informal organization. They become more disciplined in their process, while maintaining a flat structure with members at equal levels. These new breed of groups form hubs in organizations and when coupled with Collaborative Design techniques and digital communications, they accelerate the design and development process.

7. Hubs: the New Centers

Hubs are the informal centers of an organization. They are empowered by the existing social networks in such a way that they can leapfrog over the formal management structure.

John Kotter, an author and authority on leadership and change, emphasized that all organizations have "dual operating systems," with a formal management hierarchy and the informal hubs interacting side-by-side. In this way, there are multiple structures that organize and perform in a business: collaborative hubs that handle the work, and the management hierarchy that oversees budgets and allocations.[16]

The multi-disciplinary composition of a hub is one of its strengths. While one hub may focus on delivering services to a set of customers, another may customize products and services to increase their appeal in the marketplace. Hubs employ lateral processes to solve problems and can more effectively handle uncertainty in the external environments. Organizational theorist and management consultant Jay Galbraith, who developed the Star Model of organizational design, found that lateral processes often arise spontaneously to accomplish the work without going through management channels.[17] The resulting lateral relations establish liaison roles, form internal cliques, and may use direct contact with managers that share a common issue in the workflow. Decisions within hubs move priority decisions further down in the formal organization for expediency and more effective results.[18]

In today's work environments, hubs are increasingly important and more digitally connected, using group communications platforms that span networks and mobile devices. Hubs have become a vehicle for engagement for organizations that experience any number of situations:

- increased uncertainty in markets and external environments
- disruptions of new technologies
- continued pressure of global competition
- the remote nature of the workforce
- geographically dispersed offices
- the complex nature of the work itself

Hubs typically have facilitator roles at their centers, with shifting collaborative leads based on the task or topic at hand. For example, in developing software, business experts can facilitate the analysis of new features while technical specialists shed light on the intricacies of data storage, security, and interfaces. Understanding the dynamics at work in modern business and the unique role hubs play can help top management facilitate their business strategy.

Microsoft's CEO Satya Nadella emphasizes the importance of innovation and the need to build this into the culture. He readily admits that the culture of Microsoft has made mistakes in the past, most notably, its Windows Phone operating system and its acquisition of Nokia (which ended in a $4.6 billion write-off). Both were decisions made by former CEO Steve Ballmer who focused most of his energy on Microsoft's bottom line. By contrast, Nadella places much more emphasis on innovation and less on the bottom line. In his view, innovation can blossom into a something that makes money. "To me," says Nadella, "that is perhaps the big culture change—recognizing innovation and fostering its growth. It's not going to come because of an org chart or the organizational boundaries... I think what people have to own is an innovation agenda, and everything is shared in terms of the implementation." [19]

In the five years since Nadella took over, Microsoft's share price has quadrupled.[20]

Hubs can come up with new ideas, improve the workplace, prototype new products and services, or enhance efficiency of departments. There is a great deal of flexibility in how hubs operate. Yet, to be effective, hubs should have:

- Clear communications between all team members.

- Flat team structure where collaborative leadership is the norm.

- A safe environment that encourages participation, respect of opinions, and honest interactions.

- Customized cross-platform systems that facilitate interaction between members and share resources.

What makes the hub unique is its bottom-up approach. Members know the work of the organization and its customers' needs. They form collaborative infrastructures with their own unique workflows over time.[21]

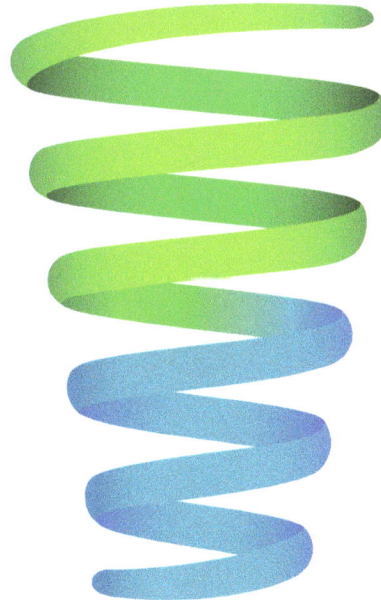

As a collective inside an organization with a mandate, a hub can communicate urgency and a working relationship that engages major segments of the organization by going beyond the usual strategy team or cross-departmental committee. They have the ability to organize quickly and the informal authority to act. A hub forces the organization to be *nimble*, *mobile*, and *flipped*.

- **Nimble** because business environments, workplace requirements, and customer expectations are all rapidly changing, not to mention core technologies. To sit through pointless planning meetings where nothing gets decided and no one takes the initiative is already waving the white flag.

- **Mobile** because teams and customers are no longer in the same place or time zone, and the organization must be available to them 24/7 no matter where they are. Whether on a desktop, tablet, or phone, teams need to interact and engage the customer at all times. Phone calls and e-mail are fine but don't always advance the ball forward on the field.

- **Flipped** because the classic top-down hierarchy approach fails in changing and uncertain environments. Cohesive teams with an intimate knowledge of the issues are needed. Their flat structure fosters collaboration and innovation.

Connecting hubs to the formal organization is a key area that empowers the decisions they make. To accomplish this, advance teams are part of the hub composition.

Advance teams

Sometimes it takes a gorilla to clear a path through the jungle thickets that can impede an organization. Not the usual kind of gorilla that leaves a path of trampled trees in its path, but one who gets everyone's attention with an urgency of a pending uprooting. A gentle gorilla.

The jungle in this case can be a critical issue, a complex problem, or a goal that the enterprise must address. By choosing a single theme to begin with, and sticking to it, management can set the tone that this topic is important. The expectation is that people in the organization (not just top management) should be thinking about it. The organization's refinement of this topic is important to the company's future.

But instead of composing an executive team that huddles in a conference room, a more expeditious step is rolling out a hub under the auspices of an *advance team*. The advance team is the gentle gorilla.

Based on John Kotter's use of *guiding coalitions* within organizations[22], the advance team is assembled from a mix of management and lower-level employees into a single team. An advance team is different than most management-appointed teams in that they act as the informal guides to form and direct hubs. Composed of a cross-section of departments and levels, an advance team is meant to set the goals for hubs to tackle. Usually the advance team takes on a specific set of issues, and begins to reach out to collaborative networks already in place. The objective is to have these networks consolidate to form one or more hubs to take on the solution. For the purpose of system solutions, these become the *design hubs*. For example, if the organization must handle a new disruptive technology that is seen as a threat to its product line, the hub may start with a central core of technical staff. But as the issue moves into product offering and customer needs, the sales and marketing experts may temporarily assume a leadership role.

Consider the following scenario. There is a continued loss of prospective new customers caused by a delay in processing credit applications. The advance team can form around the symptoms of the issue, and begin consulting with the various members of the organization that handle the processing. If the application software is determined to be a major obstacle to more efficient processing, these individuals can be brought together to form a hub to investigate and design alternatives using collaborative design techniques (described in Chapter 9).

Empowering hubs

Hubs are identified or assembled in several ways. These are usually under the auspices of the advance team to initiate a new goal that forms the hubs. There are different ways hubs can be assembled.

- **Emergent**. The member teams are in some way already engaged in the process and form more or less spontaneously. This is not always known by the rest of the organization, so there are ways to map the networks along the core theme or project.

- **Identified**. The teams can be identified by their co-workers as having specialized expertise, or experience conducive to the mandates. They can be brought into the hub for a specific goal.

- **Volunteer**. Team members volunteer to participate in the hub based on their interest.

Hubs immerse their members in a flat structure to enhance innovation. They are linked by technology which allows teams to interact more freely and rapidly than in the more traditional communications channels.

The key to empowering hubs is the advance team. The team straddles the formal-informal parts of the enterprise, clearing the way for hubs to meet, organize themselves, and begin working on the issues.

Strengthening hubs

This is not to say that hubs are perfect. They have some of the same problems any group has:

- Certain members may dominate the interactions. They are used to having their ideas adopted and may find it hard to listen and support others.

- Hubs may lose focus. They may start out with one mandate and get sidetracked into solving a related problem or pursuing a different set of objectives.

- Personal conflicts may obstruct clear deliberations and inhibit innovation of the group.

- Some of the members may fail to deliver their part, thus impeding progress.

- Because hub participation may be in addition to an individual's regular duties and workload, it can be difficult to keep on schedule.

Because of its informal nature and its cross-disciplinary composition, there is not always a single lead. Rather, the hub may have collaborative leads, or members who jump in when their area of expertise becomes central but otherwise remain in the background.

Once hubs are identified, they may need the right environment and or help in keeping their interactions collaborative. Outside facilitators or trained HR representatives can be brought in to lay ground rules for interaction and can help develop the leadership.

Types of hubs

By their very nature, hubs are way to address critical issues and strategies in an organization in an efficient way. They depend both on the dynamics of the group and the underlying technology to link resources in the company to the hub that is aimed at the right topic. From the outset, it is not always clear how the hub will focus on the higher priority topics on hand. They may start out as strategic hubs, organized around getting major business changes in place, yet during the process run into obstacles that require more focus. Unlike the conventional approach of a task group, there is a good deal of flexibility within the hub to take up issues.

In 2014, *The New York Times* created a strategic committee tasked with changing the business direction of the newspaper from a subscription-based model to a digital media company capable of handling the emergence of video content and web technologies. As part of their process, they published an internal Innovation Report[23], which was leaked to the public. In it, the committee found obstacles that stood in the way of any successful change. For example, the mindset of many of editors and journalists was to perfect their stories. The report found the following: "We must push back against our perfectionist impulses. Though our journalism always needs to be polished, our other efforts can have some rough edges as we look for new ways to reach our readers."

In the case of this internal committee, issues still lingered in the organization, ones that threaten the survival of the *Times*. In 2017, a new report was completed that describes the continued short-comings of the organization and the necessity of major restructuring of the newsroom.[24] The *Times* must be willing to experiment more in terms of how it presents its content.

Unlike committees that are involved in writing reports, hubs have more leeway to explore a topic. They are constantly seeking feedback from a larger swath of the enterprise and can organize results based on consensus.

There are several types of hubs that can be empowered within a company. In many cases, the hubs are formed based on issues observed internally or changes in the external business environment.

- **Strategic hubs.** When the business environment dictates a shift in strategy, hubs can be formed to develop and implement strategic business objectives. Typically, relevant facts and statistics are assembled to keep the hubs informed of the trends in the industry. They can point to established and emerging competition, new technologies that might impact business, and feedback from existing and potential customers. The mandate is to assemble prototypes of new strategies. These prototypes can be tested with the intended audience. They can be expanded to include other members or other hubs which are critical to the rollout of a given strategy. In this way, businesses can have a more innovative approach.

- **Design hubs.** Where new products and services are the focus, hubs can use the collaborative design approach to focus on the issues of current products and brainstorm the new ideas for products or services. The hubs can bring their ideas to a wider organization for inclusion of end users. Once the ideas are reviewed, the top contenders can be prototyped for refinement. Often these are simple mock-ups, but they can also prototype software that is run through its paces.

- **Innovation hubs**. To extend the organization's reach beyond its current set of products or services, hubs can be given a wider berth to come up with new ideas. These ideas do not need to be rooted in the existing mandates of the business, but allow hub members to think out of the box.

- **Work-life hubs.** Issues arise in organizations when the nonstop pressure of getting a product to market, or in servicing current customers, push staff beyond their typical hours. An occasional overtime can be accommodated. But as many Silicon Valley businesses have shown, the typical 9-to-5 workday may be hard to find. Of course, businesses allow their staff to take time to attend to personal matters when they need to. That said, the pressures and inconveniences of the modern workforce and its demands can create issues. These issues are best addressed through hubs, which can involve a wider audience. Many issues that cause conflict or undermine morale can be identified and discussed. In the process, key issues can be refined and serve as a focus for brainstorming solutions. These two basic processes are separate.

- **Hub of hubs.** Of course, there may be multiple sets of hubs active in an organization at any one time. Establishing a hub of hubs helps to identify each and organize them. Members of this hub may be selected by individual hubs to represent them. This allows for ongoing but flexible feedback on the progress of each.

In each case, hubs are given direction from their advance teams who provide additional resources as needed and can monitor the progress of the hub.

Final thoughts

Informal hubs within an organization create innovation and connection between teams and across hierarchy. They utilize a bottom-up approach for a task, and operate outside of the traditional, rigid org chart. By bringing together employees from different departments and with different expertise to work together, hubs can cut through static and bureaucratic red tape to effectively and creatively solve problems.

8. Digital Hubs

Hubs span physical locations of organizations and spread globally in different time zones. As conversations need to be more organized, the need for digital connection follows. Traditionally e-mail, phone calls, and text have been used with success. But as conversations become more interactive with disparate groups, new communications platforms have been introduced, including Slack, Microsoft Teams, Google Hangouts, and more.

These platforms also address the need for real-time communication and one-on-one conversations, such as Zoom, FaceTime, and Skype. Selecting the right platform depends on the needs of the organization, and the extent that existing connections need to be enhanced.

Connecting digitally

Whether you are located on-site at the offices of a company or remotely, you will always have access to either your mobile device or computer. Regardless of the time zone or location, you can engage in conversations with your hub using the available communications platforms. These provide forms of engagement and connection for hubs. There are a growing number of ways for hubs to connect.

- **Channels**. Instead of the free-form threads that texting provides, platforms like Slack organize topics, projects, and groups into separate channels. Contributing to a channel is based on the permissions given to members. By keeping conversations related to a specific topic in one place, members can view the progression of the contributions and engage along the lines of their expertise or interest.

- **Video conferencing**. As real-time meetings (especially in the design process) become more necessary, platforms provide on-screen conferencing via the user's web cam.   Video conferencing can be built into the channels or provided separately.

- **Private channels**. When conversations require privacy, platforms need to allow participants to set up their own privacy controls. This prevents other members from listening in or joining more sensitive conversations.

- **One-on-one conversations**. Many platforms allow for one-on-one conversations initiated by one member through a real-time invitation.

- **Direct messaging.** For real-time conversations, platforms provide the ability to exchange direct messages between members. The messages are housed on the platform and can replace text messages through their notifications.

- **Voice Messages**. Many platforms include voice messages, where participants can record their message and store it on the platform. Discord is one such platform, popular with gamers, that allows its members to record their messages.

- **Searches**. Platforms should allow full-text searches of previous conversations by keyword. Some platforms will limit the number of messages that are searched, based on a "freemium" model.

- **Alerts**. When messages are added to conversations, platforms will allow notifications to be sent to those participants who choose to be notified.

Once hubs are connected through a communications platform, they become *digital hubs*. Discussions, documents, and materials can be stored within channels, so that participants have easy access to the relevant information they may need.

Relationships strengthened

Digital interactions serve to strengthen the relationships of the participants, especially when groups form around common interests.

- Researcher Katelyn McKenna, alongside her team at NYU, found that online relationships formed more easily, and had a stronger success rate in establishing lasting, closer relationships than even face-to-face communications. One of the essential factors was a safe environment for exchange. [25]

- These online relationships also showed the ability to stay in place, lasting up to two years after initial contact, though they did not replace face-to-face interactions.

- Scientist and research engineer Burr Settles and cognitive scientist Steven Dow took this one step further in showing that online collaboration had a high degree of success when (1) the teams shared common interests but different disciplines, and (2) when the status between the team members was relatively close.[26] Their work also supports the idea that frequent interactions are a general predictor of successful collaboration.

- Computer scientist Moira Burke, working with Settles, also showed that online teams provide valuable motivating factors to achieve goals.[27] When used in conjunction with team learning, the social aspects of online teams can provide an additional impetus.

Communication platforms

The major factors that influence which communication platforms to use include:

- **Security.** Participants must be validated in order to use the platform, and the security of the platform must be protected. Some communications platforms can be hosted on-site behind firewalls to prevent access by unknown parties.

- **Two-factor authentication (2FA).** To ensure that passwords are not compromised from remote sites, two-factor logins authenticate the user by texting passcodes to their phones, or e-mailed prior to each session.

- **Encryption.** Both real-time messages and histories are encrypted. For real-time messaging, this means that the conversations leave the app in an encrypted format and are decrypted when they arrive at the central server. When stored or sent to other participants' devices, the messages are again encrypted.

- **Pricing.** The pricing of the platform is typically based on the number of users. Some platforms are included in suites of products, e.g., Google Hangout is included in Google's business G-Suite.

- **Data Backup.** To guard against loss of information, the platforms should provide incremental and full backups of conversations.

- **Organization**. Platforms have the ability to organize conversations into folders and links. This allows searches to be directed at the full set of both current and archived conversations.

- **Integration**. The conversations can be linked or integrated into third-party applications.

- **Languages**. Where global conversations are anticipated, the platforms will permit foreign languages and localization of their interfaces.

There are a number of factors to consider when selecting a communications platform. Many include additional functions such as surveys, calendar reminders, and member notes.

Final thoughts

Research has shown that digital hubs provide an effective means of communicating, facilitating new ideas, and keeping record of information. Selecting the appropriate platform is dependent upon both the needs of the company and its end users. There are numerous ways to connect teams and people digitally, in an array of mediums, including voice, video, and the more traditional texts and direct messages.

Part III: Collaborative Design

9. Participative Approach

To a large extent, the process used in the design creation will determine the success of the final product. Even with the right mix of participants, the communications between individuals and teams can fail to understand the problems at hand, or come up with a very limited solution. The structure and interactions of the design participants is critical: if one person monopolizes conversations and dictates the topics, the final product will be a weak design destined to be a dusty-shelf dud with limited appeal.

It's not simply the collaborative nature of the team's interactions. The steps that are used to put together a lasting design must have structure. One cannot introduce an application into an organization that fails to resolve some of the obstacles slowing down the work or negatively impacting the quality of the final products or services. To design a series of loan application screens, for example, that leave off details of the underlying security would be a step backwards. The application would be filled out by one person and when it came time to handle the underwriting (the assessment of the security), staff would need to go back to the original application to pick up the information and write it in.

In addition to having the right people involved, collaborative interactions should pervade the meetings and conversations. In this way, collaborative design engages the participants, gives them a platform to share their ideas, structures the steps they need to take, and moves the team in the right direction.

Design framework

Collaborative design is human-centered design (HCD) and structured so that each target of the digital transformation is carefully identified and examined by the design team before solutions are brainstormed and prototyped.

It incorporates frameworks used in *design thinking*, *organizational development*, and *spiral iteration*. By combining the best of these approaches, collaborative design for digital systems keeps the focus on the end product by iterating through seven basic steps, explained below. They can also be utilized by virtual teams.

The steps are:

- **Identify**. The targets of the process are identified and prioritized. These may be issues or areas that need to be made more efficient. If product offerings are included, the targets may be new features in much needed areas, or functions that the competition have implemented.

- **Refine**. Each of the targets is then further refined by those participants with the most knowledge of the area. Other participants contribute to reframe the targets in a different perspective. Wherever possible, this step drills down into the details of the target. If the target is an issue, then the presenting problem is investigated and measured wherever possible.

- **Brainstorm**. Once the targets have been refined, the team brainstorms all possible solutions and suggestions. All members participate and the goal is to list as many ideas as possible with enough detail that they can be evaluated.

- **Select**. The team then walks through the suggestions with participants explaining the pros and cons, and delving into how the suggestion can be applied or implemented. The team selects the top suggestions.

- **Plan**. From the top selections, the team prioritizes the items based on the relative risk. From this, a plan of action is put together with concrete steps on how to begin and outlines of the final implementation.

- **Build**. For each selection, members of the team assemble prototypes first, then actual production-ready components. With low-code/no-code tools, the prototypes can be coded up to demonstrate the organization, sequence, presentation, and results of the item.

- **Test**. The prototypes undergo testing and review by the future end users of the solution. Any corrections or changes are handed back to the build team, and a new prototype or component is constructed.

The process is iterative, so as one target is completed, the next is started. If the prototypes end up with issues that cannot be resolved by further modifications, the target is sent back for further review and brainstorming. The process of designing solutions relies heavily on the user and their perception of the issues that can be mitigated with the right solution. The focus on the person and the iterative process of drilling down into the issues, brainstorming solutions, and prototyping the resulting software has been used in human-centered development and design thinking. The approach builds on the model used by Edgar Schein at MIT to resolve problems and form action plans.[28] It follows the tenets of human-centered development[29] and design thinking as summarized by Tim Brown, CEO of IDEO: "Design thinking is a human-centered approach to innovation that draws from the designer's toolkit to integrate the needs of people, the possibilities of technology, and the requirements for business success." [30]

The basic steps involved in collaborative design as used for systems are straightforward:

1. **Fully understand the problem**. Many times the problems seen by the members of the design team are understood at a high level, but not in the level of detail needed to pinpoint the prime causes of stress or inefficiencies. Staying with the problem and drilling down is the critical first step.

2. **Involve the users.** As part of the process, involve the end users from the beginning to both detail the issues and to assist with the solutions and prototype evaluations. Their participation increases both the level of satisfaction with the results and how well the solution fits their needs.

3. **Brainstorm a wide range of possible solutions**. There's not just one way to approach solving the issues. Many exist, and often they don't involve developing a software solution. Again, the team delves into their creative side to come up with all the possibilities. It requires that members suspend their own judgments and perceptions to give the team the opportunity to express all available options.

4. **Select and prioritize**. From the list of possible solutions, the top ones are selected. These are prioritized and risks are identified.

5. **Prototype the top solutions**. Once the main solutions are identified, it is time to put together mock-ups, or prototypes. Sometimes these prototypes are just drawings or screens; other times they are prototypes of portions of the solution. LCNC allows quick prototyping so it fits into the design thinking approach. These prototypes are reviewed and tested.

6. **Rollout**. Finally, once the prototypes have been tested and yield the right results, it's time to implement the solution. These can be rolled out in phases, with the end users testing them in a development environment before putting them into production.

These steps take not only a large degree of creativity but also a team that works well together. The iterative process perfects a team's analytic skills, while brainstorming gives a platform for each member's creative ideas. Most members of the team know the problems first hand, but they also must be able to put themselves into the shoes of the other team members and staff outside of the team. The involvement of end users in the development process is central to the practice of collaborative design, as they both guide the team in assessing the issues while helping to streamline the solutions in a way that fits their workflow.[31]

Spiral iteration

The process of iteration and prototyping in collaborative design was refined by distinguished software engineer Barry Boehm into a development spiral where requirements, development, and testing are run through multiple times.[32]

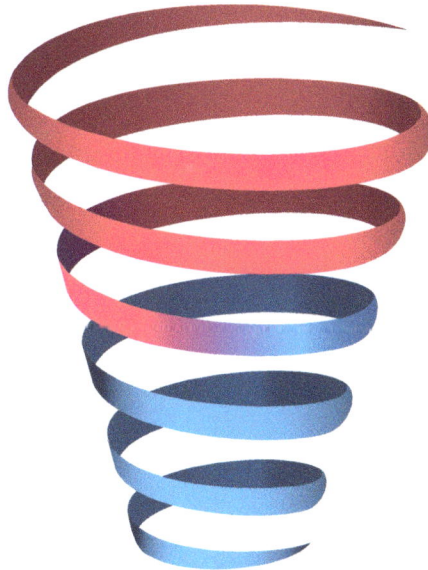

The difference from other iterative approaches is that the order of the prototypes starts with the more complicated parts of the solutions—the areas where there is more risk. The thinking behind this is that it is best way to make sure a design can handle all of the functionality. Finding out that there are significant changes to the design at the end may mean that large portions of the solution must be redesigned or recoded. Handling these areas first tests the design in areas of the most concern.

Organizational development

Collaborative design rests on the seminal work in organizational development (OD) as pioneered by Kurt Lewin at MIT, Richard Beckhard, Newton Marguiles at UC Irvine, and Anthony Raia at UCLA.[33]

These behavioral scientists focused on how to implement change in larger organizations. They approached it from an action research standpoint, measuring the effectiveness of their process on the organization. These studies were then published in peer-reviewed journals and research papers to refine their OD techniques. OD efforts assembled groups to examine and change how organizations do their work, their internal culture, and methodology to change it using people-centered processes, in response to a growing set of forces at work in organizations: [34]

- **Increased accountability**. Teams and managers take on increased accountability for the results of their work. If one team delivers late, all of the interdependent teams will also be delayed. If a team delivers poor work, their work would be incorporated into the work of others. For these reasons, managers are held accountable to deliver their products and services on time, under budget, and with a certain standard of quality.

- **Tightened allocation of resources**. The amount of resources within an organization is monitored and tightly allocated, ensuring they stay within budgetary goals. Resources beyond these allocations are more difficult to obtain, and place pressure on the teams to stay within their budgets and allocations.

- **Increased complexity of relationships**. Work has become interdisciplinary and with it, interdependent. Groups and teams must interact across a wide spectrum of departments. It has become a complex web of decision making, happening across all levels within an organization and even from external sources, e.g., government bodies.

- **Multiple goals**. There isn't a single set of goals that an organization tries to achieve, rather there are multiple sets established from different managerial levels, assorted stakeholders, and internal management philosophies. Balancing these goals puts additional stress on the workforce and their teams.

- **Impersonal systems**. People within organizations become more alienated as systems are stretched beyond the human scale. Automation reduces human activity to the entry of data and quality of life takes a backseat.

These forces, described as early as the 1960s, have gained momentum and the result is increased pressure on the people that comprise the organization. OD focused on the human side as it copes with increased demands.

One of the early behavioral scientists in the field, Wendell L. French, defined OD as:

> ... *a long-range effort to improve an organization's problem-solving capabilities and its ability to cope with changes in its external environment with the help of external or internal behavior-scientist consultants, or change agents, as they are sometimes called.* [35]

With an emphasis on group dynamics and team building, OD is able to encourage cohesiveness and reduce the personal friction, allowing the work at hand to grab center stage.

Consultants

The role of consultants, or change agents, can be central in guiding the group through a series of exercises, including iterative problem refinement, brainstorming solutions, and forming action plans.

Although collaboration is central to the success of the design process, it is not always possible to achieve it when members of the team have ongoing conflicts. This can be simply how one member performs their work and the control they need to do it correctly. For example, a scientist in a development group has ongoing conflicts with his supervisor. From the scientist's point of view, their supervisor changes the priorities of the work, interferes with the tasks at hand, and checks in too frequently on progress. From the supervisor's standpoint, the scientist continually hides what he is working on, makes decisions without consulting the supervisor, and forces him to constantly check up on the scientist.

OD approaches these situations by working with the participants, either individually or together, to encourage their collaboration. For the situations where ongoing conflicts may interfere, techniques such as *role negotiation* have been shown to have success. Role negotiation relies on the assumption that individuals with unresolved conflicts prefer a negotiated settlement to overcome the impasse.[36] These techniques and others are used to redirect participants to a more collaborative environment.

Roles

Different roles are important in keeping team discussions running smoothly, providing clear and validated resources, and allowing for continued growth of the underlying technology. There is overlap with teams. That is, individuals may be part of multiple teams. They maintain a single identity in the technology used; however, the groups that they belong may have different mandates or perspectives. There are no restrictions on this.

- **Facilitator**. The team's facilitator serves to aid the group's or team's interactions. Their role is to organize the team when it is first formed, to help assemble the rules and agenda informally, and to help resolve conflicts through suggestions and applications of the team rules.

- **Specialist**. Ideally, the different areas of the organization needed in the design are represented in the hub. For example, if general ledger (G/L) entries are anticipated to be included in the design, then having someone from accounting as part of the design team can address issues around the entries.

- **User.** For members who are actively involved in the workflow or targets, these participants can more fully detail the daily requirements and issues. Where customers are involved, they may be actively represented by actual customers.

- **Analyst**. The technical aspects of design can be enhanced by IT staff. The process of adding new layouts and activities is provided by its developers. The basic software model is extendable. However, to insure quality of the final product, developers must first demonstrate, test, and configure their contributions in a development area. Having IT present is important to handle issues such as security, interfaces, and compatibility.

- **Liaison**. Individuals who link hubs or teams together maintain a liaison role. They interact with multiple hubs and can keep formal parts of the organization informed of progress.

- **Integrator**. Where different hubs need guidance from management, an integrator role provides leadership to direct the course of design.

These roles provide the means for hubs and formal structures in the organization to work together. Keeping the different locations informed can be handled by individuals taking on certain roles.

As teams begin to meet either virtually or in the office environment, it is important to talk about *rules*. The rules cover how members interact. If members veer from the rules, other team members can suggest a different approach that member might try instead. If a discussion becomes more heated, a team member can ask about some the underlying facts that may have caused a more heated response. Facilitators can also make suggestions if the team seems to be stuck or conflicted. Conflicts between individuals can be handled offline by defining roles the individuals will assume at future design meetings.

At each step of the collaborative design framework, the five listed below represent the more urgent areas of discussion.

- **Identify**. The team discusses the agenda, setting top goals by identifying broadly the areas they would like to address. These can be issues, new product offerings, or features.

- **Refine**. For each of the items in the agenda, the team drills down to more detail. For larger design teams, the group may be divided into smaller teams for this step. They reassemble for discussion on each of the items.

- **Brainstorm**. The group then brainstorms solutions and suggestions. All members participate and the goal is to list as many as possible with enough detail that it can be evaluated.

- **Select**. This step begins to prioritize the suggestions or solutions, and select the ones to pursue. The facilitator should make sure participants have equal time to express their opinions. Voting can be used to reduce the suggestions to the top items.

- **Plan**. From the selected solutions, the team prioritizes items based on relative risk. A plan is put together with concrete steps on how to move forward.

These can be refined by each team but serve as a structure for meetings both online and offline. The process is iterated as problems gain more clarity and solutions become more detailed. As plans evolve, the team then begins the prototype and testing steps.

Trust

Trust is something developed through repeated interactions, but is to a large degree in place when hubs know one another outside of their design meetings. This doesn't mean the hub members need to be friends (or even colleagues), but some sort of common ground should be established.

Research into effective groups has also shown that safe spaces are also important. Teams that can share common thoughts and emotions can be more productive.[37] Although this applies to workgroups, it carries forward in teams. Keeping teams small promotes this kind of interaction.

At Google, Project Aristotle was started to find out what made teams effective. After interviewing teams, they found were three basic determinants: (1) everyone in the team felt they could speak their mind and others would listen; (2) everyone was sensitive to how others feel; and (3) everyone had equal time to talk.[38]

Final thoughts

Active participation with clearly delineated roles is key to the success of a design hub. The tenets of organizational development are meant to address existing concerns in the modern workplace environment, and provide a framework for hubs to collaborate and implement sustained good practices and outcomes.

10. What Can Go Wrong?

As supportive as the design teams can be in their Collaborative Design process, there are a number of issues and situations that can sink their efforts. Many of these, if they are recognized early, can be avoided. Others are better handled by the facilitators or consultants.

Getting sidetracked

Some topics easily lead to others, and members can get sidetracked in their focus. The first inklings of this happening can be casual conversation about related situations. Examples may be given of how some issue is linked to another, and discussion can ensue. Spending a small amount of time in discussion can be accommodated.

The best solution is to suggest that the item be written down and covered either at another session, or tabled until end of the current one.

Dead ends

In brainstorming solutions, the design team may hit some dead ends. Perhaps the ideas do not address the issues or goals at hand, or they are off course, or have been tried before, or members are not thinking outside the box. It could simply be that the group has reached their limit, especially at the end of a meeting.

One way of approaching this is reframe the issue causing members to go quiet. Instead of looking at it as a problem, for example, suggest that this doesn't need attention at the moment. Or that the roots of the issue are counterintuitive. A change of perspective or even venue can recharge the team and get brainstorming back on track.

Loafing

Designing may be an occasion for some members to take a break and listen. Although there is nothing wrong with this initially, participation by all of the design team in a collaborative framework is what is important. A good way to handle this is for the facilitator to ask each member for their opinion on certain items.

Not engaged

Other members may simply not care or be engaged in the conversations. Although it is hard to control in virtual meetings, phones and other distractions should be set aside. The facilitator again can direct questions to those participants who do not seem to be engaged; it may be that they have a problem with something in the meeting, or simply don't understand a given topic.

Holding back

When a participant isn't engaged, it may be that they have an issue with the agenda, the process, or others in the group. This is a more difficult symptom to detect and hard to handle without sidetracking discussions. One approach is to use the end of the sessions as a way to summarize the progress that has been made and also to encourage suggestions for how things can be run differently. On this topic, encourage those who have not been engaged to make some recommendations.

Missing members

There are times that members cannot make the design sessions. This is not uncommon. But when larger numbers do not show, it is cause for concern. There may be individual reasons, e.g., there are project deadlines that are getting in the way, or a scheduling conflict.

The solution is both individual (where facilitators contact these participants) and group (where other members also contact these members). Both approaches aim at figuring out why members aren't showing and addressing the practical matters as best as possible. If individual schedules are causing problems, then the group can discuss more optimal meeting times.

Too many targets

When members generate a large number of targets in their Identify step, this can overwhelm the group. It may simply be too many to try to resolve. Some may be related and combined. Others may be lower priority. Certainly the group can attempt to address them in the priority of their importance, but if the list is long, the group should then set aside time to postpone portions of these items.

Lack of management support

Since a good deal of the sessions are with the informal hubs, managers of different sections or departments may simply not want to provide support for a project outside of their direct report's given workload. This is unfortunate, as it could prevent any substantive designs from being worked on. The causes may be that the members themselves initiated the work outside of the formal chain of command, or that they went beyond the agreed upon scope. In these cases, the matter should be brought to management's attention. It is only the formal organization (i.e., managers) that can solve this matter.

The approach should be for each set of coworkers to address their higher-ups for their suggestions on how best to proceed. This may be contacting their direct managers, or it may be informally discussed with other managers. Management is largely in control of the time and resources involved in the design process, so it is critical that it has their buy-in. Many times, it is upper management that sets the design as a high priority, with levels below having to come up with the time and resources to handle it. This in itself can be a separate set of sessions to determine how to handle the situation.

No facilitators

Facilitation of the design sessions by one or more members adds to the efficiency of the process. When no one assumes this role, it can leave the team in disarray at times. Discussion moves from topic to topic, and some members can dominate. When no one person has this role, it can be rotated. For example, one member can lead discussion of the issues, while another keeps the brainstorming going, encouraging teammates to be more creative. These roles then shift.

One way to handle this is to discuss how it is everyone's responsibility to facilitate discussion. If one member dominates or another is silent, then it is up to the others to step in to facilitate a collaborative environment. This can be the first item of business when the team assembles for a design session.

No prototypes

The Build step is one of the most important parts of the process since it gives members a chance to prototype the proposed solution. For those members who want to jump onto a LCNC platform and assemble a prototype, it can be a fun and challenging activity. But in cases where no one volunteers or no prototypes are completed, the team should meet to discuss other options. Not all prototypes are coded up. Some are simply sketched out on paper. These can be the interim prototypes. But the team needs to see if others can be approached to help out.

A good solution is for team members to approach others in the organization that they know to help. They may be part of IT or have had experience with other LCNC solutions and may be able to get a prototype off the ground. Management can also be contacted to see if they have others that can help.

Final thoughts

Assembling a hub is only one part of the design process. Even with clear roles and defined steps, there can be many impediments to action, many of which can be resolved by ensuring team members feel included, supported, focused, and engaged. It is imperative that each member fulfill a role in order for the iterative design process to be successful.

Part IV: Low-Code/No-Code

11. What is Low-Code/No-Code?

Low-code/no-code (LCNC) software development is a simpler way to develop technology. It does not require extensive knowledge of coding or software design, but rather is based on the business knowledge of its developers. "Low code" development uses a combination of drag-and-drop objects and higher-level languages to develop applications. "No code" development does not require any lower-level languages and uses drag-and-drop to build systems.

When the design team begins building prototypes, one of the easiest ways to do this is to have members use LCNC platforms, as they are relatively quick to implement. These prototypes are part of the Build and Test phases of collaborative design.

As the design project moves forward, IT's involvement can be focused on the technical aspects of development, testing, and installation. IT ensures technical fit with the organization's requirements and consistent security across platforms and locations. Software engineers can still be at the forefront of LCNC coding, since they can quickly understand the platform. They can parcel out tasks to members of the user groups, such as having them construct the screens or lay out the reports.

A classic example of LCNC software at the individual level is the spreadsheet. Typically, these are assembled by users to help with tasks such as data analysis, project management, or financial projections. The first microcomputer spreadsheet, VisiCalc, was introduced on the Apple II computer in 1979, followed by Lotus 1-2-3 in 1983 on the IBM PC. Microsoft then introduced Excel in 1985 which overtook 1-2-3 as the main spreadsheet.

Microsoft Excel (part of its Office Suite) is a simple example of LCNC software. Users of spreadsheets need only understand the basics of entering formula into the individual cells of a row/column spreadsheet. Entering the basic numeric information in one column of cells, their corresponding formulae are entered into the second column. Filling the formula from the top of the column to the bottom converts each formula to use the numeric value to its left.

Spreadsheets open up a world where users handle their own software to automate many of the processes of their daily work. Financial statements can be housed in spreadsheets; projects can be tracked; and business projections can be made.

The problem with spreadsheets for hubs is that one person's data or formulae are often out of sync with another's. Changing a value in one spreadsheet does not affect the value in spreadsheets created by other members of the organization. This doesn't mean that spreadsheets cannot be used by hubs to perform necessary functions, but it means that the spreadsheets must link to one another and that there are controls over who can change information and formula on the sheets.

At a certain point, spreadsheets begin to lose their effectiveness, and databases enter the picture as central repositories of information that can be kept current.

IT has its hands full most of the time

Off-the-shelf software (i.e., packaged software) offers a number of conveniences. The staff does not have to spend time designing and developing their own software for their unit. Instead, the functions of the packaged software can be rolled in.

A problem with packaged software Is that does not always provide the flexibility or functionality needed by the variety of hubs in the company. The initial steps in considering whether to use off-the-shelf software or take on LCNC development remain the same. The functions, documents, reports, and interfaces still need to be established. From these lists, the organization can determine whether to invest in packaged software or begin an LCNC project. Chapter 12 (*Why Not Just Buy Some Software?* on page 95) walks through the process of evaluation. Chapter 19 (page 163) outlines the first steps to review available third-party packages.

It is important to understand that IT traditionally has a number of constraints when taking on new projects or adding staff to ongoing ones:

- **Departmental priorities**. The IT department of the company may have its own priorities that delay the start of a development project outside their usual purview. Urgent problems that pull technical staff away may further complicate getting on the docket. So for a hub (which may cross departmental lines) to rely primarily on IT for development may put too many eggs in one basket.

- **Maintenance**. For similar reasons, IT may resist getting involved in new systems that must be supported by their staff; IT relies heavily on its staff in place. Most IT departments have strict budgets to control their ongoing costs to the company. Software maintenance can consume 60% of the department's budget. [52]

- **Slippery slopes**. Once an IT department begins working on new software, they may prefer to use development tools that their staff is already trained on. For example, using Javascript or C++ is common for new projects as it gives the most efficient approach. These languages are well-known, so any maintenance will have IT staff already well-trained in the languages, though other members of the hub may not be.

- **Requirements**. For IT to do its job, it needs to understand the requirements fully and these requirements need to be static. To design to constantly changing requirements means a development project that never ends. Further, IT would need to understand the business and organization requirements.

- **Design time**. IT staff will also need to fully understand the scope and context of the development. These requirements will define the data needed and the relationships between the different data fields.

- **Costs.** If the project is projected to involve additional third-party software, licenses, or ongoing data, this can impact overall costs, plus any additional IT staff.

Another downside is development costs. As more lower-level programming is being used, the more coding, testing, documentation, and debugging are needed. Long-term maintenance increases as the complexity of the code increases.

IT remains critical in the development of LCNC software, since they oversee security and help with access the other systems in place. But the major development work can be offloaded to the available members of the team.

LCNC software capabilities

Low-code/no-code development offers a way around these obstacles. It does so by placing the design, development, and support of the software on the shoulders of the hubs themselves. This means that the members of the hub collaborate to develop and deploy the software they need. As requirements change, the software is updated by the members of the hub. Since hubs span departments, LCNC systems are most likely to consider the requirements across multiple disciplines.

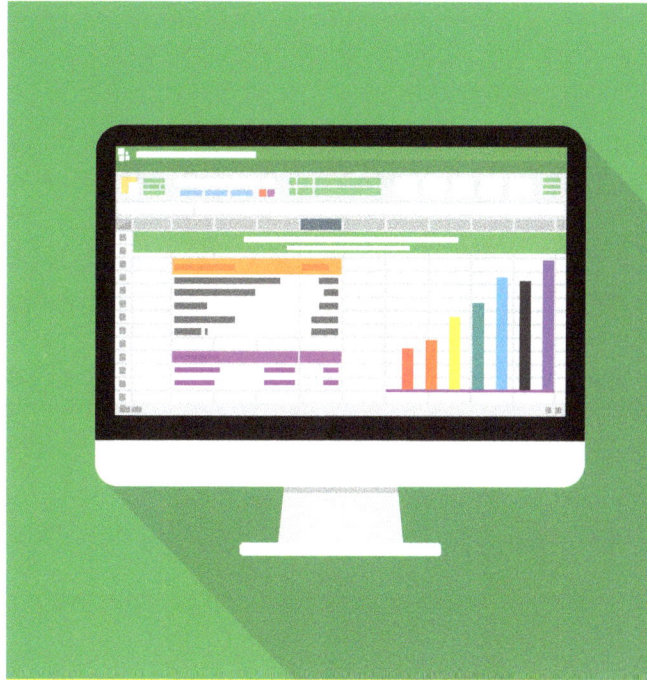

The types of software vary, but they fall into the following categories:

- **Websites.** Website construction allows a user to develop a private or public website with information about products and services. The sites can include forums, blogs, news, subscriptions, chats, instructional videos, and more. Private sites can be hosted inside an organization with password-protected information aimed at keeping coworkers, management, and customers informed.

- **Workflow**. Although package software may handle a good deal of the security of the data, including the roles needed to access or change information, LCNC can be overlaid to implement a workflow within an organization. This can be anything from multiple approvals to sending members summaries of applications; the net impact is to establish a route for information to take before it is finalized.

- **Database extensions.** The databases that form the foundation of package software are usually set in stone. This is to prevent incomplete information and to optimize performance of their code. However, organizations need to collect additional information specific to their business or goals. To do this, additional information must be collected, maintained, and integrated to the package software.

- **Cross-departmental reporting**. Most packages have a variety of reports, from financial to statistical. But many times, reports need to take different forms and have different filters. This can be implemented with a DIY report formatter that comes with the package, or purchased separately as LCNC add-ons.

- **Data modeling and analysis**. Above the statistical analysis of organization information, there may also be a requirement to model the data to be able to forecast the impact of events on the organization, or a set of hubs. The modeling and data analysis gains access to the basic information and extends it through its additional computations. For example, risk or cost analyses are very useful processes that can be an overlaid on business data.

- **Transactional**. When hubs are involved in receiving and processing transactions, the software needs to incorporate a set number of steps with some sort of approval process. Each step may involve connecting to other steps in the process. For example, a loan application may need to have additional information about the prospective customer's financial information before moving on to an evaluation phase.

Each type of business or organization has its special needs. As departmental boundaries are crossed, the approach to applications can change. Hubs having access to bring the disparate viewpoints together and construct a unifying set of application helps the effectiveness of the hubs.

Integrated functionality

Much of LCNC software comes packaged with additional functionality which can be added and customized to fit the needs of the end user. Since these components are pre-built to work with one another, they also have the advantage of providing the underlying database structures, report formats, and on-screen dashboards as needed.

- **Screen design.** There are a number of screen types that can be used and customized by dragging components to a mock-up screen. For example, to add a title to a screen, the designer selects the Text component, drags it to its place on the screen, then enters the text. The font type, color, size, and alignment can be customized. To place a picture, often the designer can simply drag the picture from a folder.

- **Form design.** When users need to fill in information on the screen, LCNC platforms provide forms that can be designed to speed up the process. In many cases, the values on the form (such as a customer's address) are already known and can be defaulted on the form. A number of field types are provided, such as dates, currency, text areas, checkboxes, and lists. The designer lays these out on the form in a way that is intuitive for the user.

- **Database storage.** As forms are built and added to the system, the corresponding storage is also created in the platform's database. Changes to fields impacts the storage of the built-in database.

- **Web design.** As part of screen design, LCNC platforms may also have a separate set of components that are targeted at web implementations. For example, adding pop-up ads or a newsletter opt-in may be available for screens displayed over the web.

- **Role-based access.** With different types of users ranging from administrators and software engineers to data entry clerks and supervisors, many LCNC packages include the ability to define levels of access. This ensures that only authorized members are allowed to change the code or handle the data at a granular level.

- **Mobile access.** Websites, forms, menus, and views have to adjust to the platform that they are on. Most LCNC platforms build this responsive capability into their screen, view, and form designs. For mobile access, the final application is aware of the device and its orientation to make its responsiveness automatic.

- **E-mail**. E-mail functionality may be included in the LCNC platform. This allows the user to send and receive e-mail without leaving the LCNC application. In addition, the designer can automatically generate e-mail, notifications, and texts as needed.

- **Views and reports**. Reporting can be done on-screen (as views) or in printed format (as reports). These can be designed by indicating which fields go into which columns, and how each is formatted, summed, sorted, grouped, and presented. Many times new fields are needed in the reports, such as for the total of a selected column. LCNC platforms allow the fields to also be stored in the database in separate collections.

- **Interfaces**. As LCNC applications are developed and rolled out in organizations, they must interface with existing systems. For example, accounting systems need an end-of-day set of accounting entries in a particular format. These interfaces can also be part of the LCNC platform, making it easier to connect applications.

- **Data warehousing**. Although the LCNC has its own database structure and access rules, many LCNC packages provide vehicles for this data to be written to larger, shared organizational data warehouses. This allows multiple LCNC platforms to be used alongside larger systems.

- **Data science and analytics**. Data warehousing provides the ability to use data science to analyze the underlying information. For example, predictive regression can be used to detect fraud, or identify churning with customer patterns.

- **Artificial intelligence (AI)**. Many LCNC platforms are AI-enabled. This functionality can make recommendations for customers based on previous purchases, or help optimize customer service. AI can also customize experiences for customers, provide suggestions to management, and highlight next steps in complex situations.

LCNC packages differ and include different sets of components. As part of the selection process, the design team lays out the functionality that it needs and evaluates the LCNC platforms on their ability to meet these requirements.

Expanded interactions

The software also lends itself to new ways to present information and to engage its audience. In addition to the standard business-oriented screens, LCNC code can be expanded with more creative User Interface (UI) or User Experience (UX) designs.

- **Information overlays.** To accomplish this, information overlays span multiple systems, provide rudimentary analysis, and give teams both visual representations and access to the supporting details. Individuals can exchange items and information with others on their team. Embedded in the platform are notes relating to various topics. The notes are connected to the source material through links.
- **Visualization**. This is a visual 2D representation of the organization, composed of locations, teams, and people at various levels.
- **Idea management**. This software provides a tool for crowdsourcing and generating new ideas and suggestions, while also building engagement among team members.
- **Challenges**. Once teams move to higher levels, they are ready to be challenged by more pressing questions. These questions aim at solving important issues and are used by the team to craft solutions (which can be voted on by the rest of the organization).
- **Messaging and chats**. Individuals interact in a separate messaging system (one-on-one) or in chats (full team). These interactions are private to the organizations. Individuals can also use other products (e.g., Slack) for real-time messaging.
- **Leaderboards, badges and points.** Teams and individuals are rewarded in various ways. The better the teams do, the higher they rise in the game.
- **Suggestions and surveys.** These are normally sponsored by the department, and allow for feedback on a wide range of topics and issues. Individuals may reply anonymously if they choose.

- **Games.** These are puzzles that team members must solve. They are given a preset amount of time each day to solve as many as they can. The puzzles are fully customizable and may be based on subjects that the organization wants employees to know more about.

Online communications

With a dispersed workforce, online communications becomes more important. Although traditional means still work, LCNC software expands these venues to include video, voice, messaging, and customized group discussions. The basic functionalities include:

- **Direct messages.** Members of the enterprise can interact one-on-one with direct messaging, much like phone-based text messaging, which are sent in real time to the recipient who can view them and respond over the platform.

- **Real-time discussions.** Group discussions can be either public or private. For public discussions, members need to be given access to the workspace (where the communications platform hosts the organization). Members can view and participate in these discussions. For private discussions, only select members are given access.

- **Video conferencing.** Members can interact with one another with video conference calls. The functionality is integrated into the platform, with permissions to allow a select group to conference or to open it up to the wider membership. Conferences can be recorded for future reference and sharing.

- **Voice conferencing.** Phone calls can be made over the internet using the platform's voice conferencing facilities. This allows groups to talk with one another. Calls can be recorded for future reference and sharing.

- **Workflows.** Many platforms provide structured workflows so that groups can walk through a series of steps as part of their interaction. For example, technical support inside an organization may walk users through a series of questions about technical issues before linking up to an IT member.

- **Bots**. Automated functions can be assembled to call when a topic or event occurs. These work to reduce repetition, lower the volume of requests on technical support employees, or to collect information.

- **Documents and files.** For ongoing work that the members do, the platforms provide easy access to documents and files. These can include reports, drafts, spreadsheets, diagrams, schedules, test results, images, video, and notes.

- **Links**. Members can also include links to other discussions, websites, and files. This helps to supplement discussions with research or examples.

- **Channels**. Channels offer a way to organize topics and groups. For example, there may be a channel on a project dedicated to suggestions.

- **Cross platform**. Platforms support most desktop and mobile platforms. This allows conversations to continue from the field.

- **Localizations**. For members located in different countries, most platforms provide localization features, for example various language versions, date and time formats, and time zone adjustments.

This type of functionality can be delivered over integrated and standalone low-code/no-code platforms.

Technical fit

There are a number of low-code/no-code platforms that allow members to build their applications more easily. In most cases, access to the underlying data is necessary and the ability to add and change information can be important. For example, if a financial instrument needs added fields to indicate risk, the platform should allow the developer to add a "risk" field to the customer's record. Or if the underlying collateral is valued based on the market trends, the designer should be able to read the necessary fields, make the computation from public information, and place this with the collateral.

Just how information is accessed, edited, and stored is a functional of its technical fit:

- **Application Programming Interfaces (APIs)**. Where members need access to the underlying data (e.g., customers or accounting), the platform should accommodate API interaction. APIs provide a range of functionality, so the platforms should be able to read from and write to the underlying business data using APIs.

- **Security**. Where most platforms take care of security (e.g., logging into the organization's network), the platform ideally uses the same security. This allows a single sign-on. However, if this is not possible, or if additional roles are required, then the LCNC platform must have this implemented in conformance with the organization's security protocol. For example, if a two-factor authentication is required, the platform should support this.

- **Industry compliance.** Within certain industries such as banking, finance, or medical, the data collected and stored must meet certain standards. This includes both the formats and storage of information.

- **Privacy**. Where platforms are hosted in the cloud, there must be a verified adherence to data privacy.

- **Reliability**. The assumption going into LCNC development is that the program is reliable. But this needs to be fully investigated, since many programs may break easily. For example, if the underlying database has mandatory fields of a certain type, but the input is not validated properly, the database can produce unrecoverable errors.

- **Ease of use.** Learning the LCNC platform is critical for both the end users who participate in the prototyping phase and for the organization's ultimate programmers. If the LCNC development interface is hard to understand or master, this may end up slowing down the application development process.

- **Data backup**. Where data is stored outside of the central database, data redundancy and backup should be implemented. Typically, this can be done with the organization's backup and restore procedures.

- **Updates**. As package software is updated, the platforms must be able to update their internal code to be in sync.

These technical requirements ensure that the LCNC applications do not run into security or data issues down the road.

Which LCNC works the best?

In choosing the right LCNC platform, there are a number of considerations that should be evaluated. Ideally the same platform can be used across hubs so that applications can be extended to seamlessly incorporate other hub requirements.

- **Price**. Most LCNC platforms price their usage based on the number of installations that use its tools. These are priced by the number of active seats or users (SaaS). The pricing must also accommodate installations where there are a large number of users, or a global customer base.

- **Compatibility**. Most platforms will indicate the operating systems, devices, cloud storage, security subsystems, and environments that they are compatible with. All of these make it easier to implement LCNC solutions across an organization's user base.

- **Templates**. Many platforms already have templates for standard processes, reports, or analyses. This should be considered when evaluating platforms. For example, instead of designing a report for outstanding issues in a customer relationship management (CRM) application, the LCNC platform may have several report templates that come standard and can be customized as needed.

- **In-house expertise**. Since it will be principally the members of the hubs that design and develop using the platform, prior experience on the platform makes its implementation easier and quicker.

- **Security**. The ability to restrict both access and viewing of data is critical in most environments. This is especially true of insulating data from outside intrusions. This may mean the intrinsic use of encryption in data storage, or the tracking of access outside the office's internal networks.

- **Ease of use**. The platform must be intuitive to use and present clean interfaces. This can be evaluated through demos by the design team. Many time the inclusion of templates help users to lay out their screens.

- **Scalability**. The LCNC platform must accommodate the number of users and locations that an organization envisions for its application deployment. Although a LCNC platform may work well in a small- or medium-sized enterprise, it may run into performance issues when implemented at a larger scale.

- **Technical fit**. The platform should be evaluated alongside the technical requirements that are needed for the platform to communicate with the underlying systems and the assortment of other applications in use at the organization.

Each platform has its own set of advantages and disadvantages. By prioritizing these criteria, organizations can better understand which platform works the best for their requirements.

Final thoughts

In selecting the right platform all business applications in use should be examined with the LCNC platform's ability to connect with them. LCNC software platforms give added control back to hub members while removing the burden of a heavy workload from IT departments and proprietary software development. LCNC functionalities are especially useful for those members who may not have expertise or a technical background in platform creation.

12. Why Not Just Buy Some Software?

The obvious question is why not purchase off-the-shelf software or subscribe to a software service (SaaS) to address the issues? In many ways this can be an excellent solution. The cost is known up-front, and a support staff is available from the company to handle any questions or bugs.

Since a list of documents and reports has already been collected, and the necessary steps of the process identified, an organization is in a perfect position to see if off-the-shelf software makes sense. This is usually where the IT department can get involved for their input. Since there may be existing systems in place, IT has a better understanding of the interfaces and the ability of systems to share data.

List of systems

To this end, IT can also assemble a list of systems in use by the organization. This can be used to determine what interfaces are needed in off-the-shelf software and also what migration is needed if a new system takes over portions of the processing from other systems.

System List Example

Software Used	Purpose	Comments
Constant Contact	For e-mail campaigns for marketing.	
Contacts Plus	Houses both prospective and existing customer information in a secured database	
Jack Henry Accounting	Accounting system that handles financial reporting	

In each case, the system is used for part of the overall process. There may be an off-the-shelf package that has interfaces to these packages. If so, there should also be a migration path when a new package is installed. This brings over the data.

How customized do you need it?

One of the determining factors is the amount of customization that is anticipated to fit your environment, customers, and possible future changes. Off-the-shelf packages may in fact be able to handle these needs. For example, if there are changes to industry requirements, most off-the-shelf packages are set up to implement these changes. By contrast, if a LCNC program is preferred, an organization may find libraries or plugins that do the same things.

What's an RFP?

One of the best ways to find out if an organization's needs match the features of off-the-shelf package is to prepare a Request for Proposal (RFP). This outlines the types of tasks, data, documents, reports, web access, and security needed by the package. It can also describe the technology in use at the organization, for example, Windows desktops or Apple mobile devices.

The RFP takes time to prepare, but has the advantage of getting a complete quote from vendors, including installation, training, and ongoing maintenance. It encourages vendors to give the best price since they are being compared to other products and companies on the market.

Let's see it!

Once quotes have been received, an organization can call in the vendors and present to the hubs. The more people that preview the software and understand its potential benefits, the better the evaluation. Staff can see how easy it is to work with the screens and generate the needed documents. IT can see how well the data storage and technology fit. Management can review the reports and costs.

Final thoughts

LCNC platforms provide for accessible software and customizations tailored to the needs of a company or customer, but sometimes these needs can be met just as easily with readily available, off-the-shelf software. Similarly, companies can license or subscribe to existing software, in the form of SaaS platforms.

13. Low-Code/No-Code Platforms

There are a number of well-known platforms that are used in both the initial prototypes of designs as well as the final digital solution. The following list can change over time. Since there are different types of LCNC platforms, it is best to first determine which general approach would work, then to look through the list of available LCNC platforms.

Business process management (BPM)

For business processes, LCNC platforms provide the foundation for basic workflow with business data (such as orders, payments, staffing, and events) stored in secured databases. Portals for customers can be developed with strict access protocols. Drag-and-drop functionality with minimal code makes use of the platforms efficient. Costs depend on the vendors. Data storage can be in the cloud or on-site.

Examples of this type of LCNC software are:

- **Appian**. The platform builds business processes with drag and drop, incorporating business rules and policies and allowing users to track and analyze markets, events, and other changes. Forms, dashboards, and workflows are part of the development process.

- **HCL Domino/HCL Notes/HCL Domino Volt**. The HCL platform uses forms and views to construct basic business components for workflow, inventory, events, payments, and scheduling, with integrated email and layered security. HCL Notes is the client-side software. HCL Volt is a newer low-code tool that allows design teams to build multi-platform apps with drag and drop components. Templates are available for finance, sales, and marketing and customer service.

- **Microsoft PowerApps**. This platform uses basic no-code functions to build processes around tasks and roles. It incorporates location tracking and camera input in its basic functions. Users identify the data fields and the basic workflow to form the business model. Web portals allow customers to access specific portions of the applications.

- **Mendix**. This platform builds the basics of design into its process, where users can identify design team members and walk through possible solutions. Data logic is organized and screens are put together with drag-and-drop components. The platform has built-in interfaces to other third-party applications.

- **Zoho Creator**. This platform allows the design team to drag and drop fields to build forms, dashboards, and perform analytics. There are a number of built-in functions to handle event scheduling, workflows, orders, payments, and inventory. Templates are available for education, manufacturing, finance, sales and marketing, HR, and customer service.

These platforms can also assist in the design process by providing structured steps in the development process.

Customer relationship management (CRM)

To handle the marketing to prospective customers and the ongoing relationship between a company's customers and their internal staff, customer relationship management (CRM) software streamlines contact management, the creation and running of marketing campaigns, and the analysis of customer interactions.

Examples of these are:

- **Salesforce.** This product integrates e-mail, marketing campaigns, live chats, and third-party APIs into its services. Contact and document management are included to help keep customer files organized.

- **Microsoft Dynamics.** Dynamics runs multichannel marketing campaigns with access to LinkedIn to generate leads. It integrates with Office 365 applications over Microsoft's cloud platform. AI is also incorporated to inform business decisions.

- **Monday**. This work OS platform is a digital workspace that uses project management structure and visual dashboards to monitor customer interactions.

- **Freshworks**. Freshworks incorporates customer service with its sales and marketing functions. Contact management is included with a de-duplication feature. SMS can be integrated into the platform.

Interfaces between these CRM platforms and other systems (typically through APIs) allow basic functionality to be customized by the type of product or services being offered.

Outward-facing websites

To reach customers or to inform the public, often websites can be constructed with little or no coding. Some of the platforms can be integrated with BPM software, e-commerce, and customer relationship management (CRM) software.

Examples of this type of platform are:

- **Weebly**. This web development tool provides easy site construction with no code. The site is responsive and can be accessed over all devices. SEO and analytics are provided.

- **Wix**. Requiring little development background, the Wix platform provides a collection of over 500 templates that can be customized for outward-facing sites. The sites can be accessed over desktop and mobile devices, with responsive formatting of screens. SEO and analytics are provided.

- **Wordpress**. With a vast collection of plug-ins and themes, Wordpress gives users the ability to drag and drop fields, buttons, graphics, titles, animations, security, and backgrounds. Little coding is necessary. Data is stored in Wordpress databases and can be pushed to back-office systems. SEO and analytics can be added.

- **Bubble**. Bubble is a cloud-based hosting environment offering drag-and-drop features that allow users to build their websites with integrated workflow. Information can be stored from the website to a stand-along database, and solutions can be scaled up.

Other web platforms are more specialized, such as Shopify for online retail services.

Online group communications

Some of the best tools for engaging teams online are available through a number of communications platforms. These applications take the basic messaging and forum structure, and expand them into tools that can be used for online discussions.

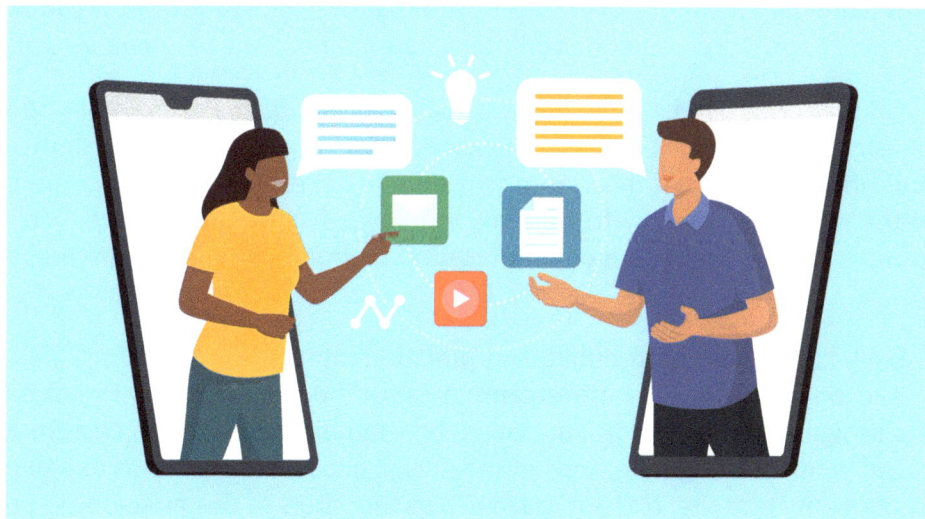

Typically, both private and group communications are handled. Examples of communications platforms include:

- **Slack.** Members of the enterprise are invited to workspaces to discuss topics of their choice. Each topic is a channel, with both public and private options. Direct messaging, notifications, integration with cloud storage, and the ability to design online workflows are included.

- **Microsoft Teams**. Corporate environments are created with both small and large group conversations. Teams integrates with other Microsoft Office products (e.g., Word, Excel, and PowerPoint) as well as with Skype for online meetings.

- **Google Hangouts.** Hangouts is designed for both personal and work-related groups with integrated voice, video, and instant messaging. The service can link to Google Sheets, Docs, and other products.

- **Chanty.** Chanty is a simple team chat application that supports both public and private discussions. Links, files, and conversations can be organized in its Teambook feature.

- **Flock.** This application provides threaded conversations along with audio, video calls, reminders, and notes. It permits organizations to connect additional applications and supports screen sharing and searchable message histories.

These platforms allow for customization using LCNC plugins. Ongoing workflow can be augmented with these products to give organizations additional communications between its staff.

In addition, LCNC platforms can be used in the design process itself. For example, to structure the collaborative design process, organizations can expand their engagement through the use of online group communications. In the past this was mostly handled in e-mails, draft documents, and meetings. But with the expansion of work-from-home employees and remote offices, organizing these under the umbrella of a central group communication platform can facilitate a more efficient approach.

For example, Slack can handle both public and private communications, direct messaging, video conferencing, and phone calls. They are organized into different channels, based on the subject which can be a design project. In addition, discussions can be structured as custom workflows, so that members are guided through the steps of the design process. Documents and diagrams can be linked to the Slack conversations. This approach allows members of design teams in other locations or that work remotely to participate in the collaborative design process. It also provides a way to organize supporting documents, references and graphics.

Final thoughts

Selecting the right platform depends on the type of system that is being considered and the needs of members and users. Some platforms may be too pricey while others do not have the security or scalability that is needed. Part of the collaborative design process involves building prototypes on different LCNC platforms. It is a perfect opportunity to try out a platform without a heavy investment in subscriptions or training.

Part V: Digital Design Hubs

14. Bringing It All Together

On their own, LCNC platforms and collaborative design provide major boosts to the speed and quality of software solutions. When paired with design hubs within organizations, they gain traction since the teams are mobilized and empowered to move forward. The approach to maximizing the effort and keeping continuity across a dispersed workforce and disparate issues is to connect hubs online so that members can be in a structured design process.

Design hubs

A first step in the development effort is forming the design hub itself. This is the informal set of teams that are mobilized to handle the design and development of the software solution. Current hubs in an organization can serve as the basis of these teams since they are the closest to the work and have knowledge of the intricacies of the workflow.

They also have first-hand knowledge of the problems encountered and the exceptions that need special care. It makes sense to look closely at these hubs to begin to understand how they can be involved in the design of new LCNC software. Design hubs themselves can be formed in a number of ways:

- **Active interest.** Hubs themselves may form to explore better ways of doing the work. This may be from the standpoint of reducing stress, or to optimize the efforts of the staff. Discussions may begin spontaneously, and evolve to the point where members begin exploring ways to improve the workflow through their involvement.

- **Solicited involvement.** The advance team may solicit suggestions for how to handle problems or make the workflow more efficient. This outreach can be made casually, for example at meetings or in e-mail to staff, or it can be part of an informal project. Hubs can form with a common interest in pursuing new ways of doing things.

- **Selecting a core group.** The advance team typically appoints a team to subsequently form the design hub. This core can then populate their teams with individuals they feel have first-hand experience with the work under review, or with experts that can provide insights on how best to handle the design.

- **Current and future users.** Users closest to the work know the issues and can help brainstorm solutions. Many times these users have already tried a number of software solutions, and have experience with how the workflow is best automated. With LCNC platforms, they are now able to assist in the development.[39]

- **Appointing a design team.** The advance team may simply appoint members of the design team. The disadvantage to this approach is that the design teams may not involve the right people, that the structure of the hub retains the formality of existing management hierarchy (which can inhibit the creativity and range of solutions); the appointed members may not have an interest in being a part of the design process.

The more collaborative and innovative their process, the better their results. The process of forming the design hubs begins by reaching out to the different areas involved and identifying the critical players.

Teams

The design hub itself is composed of multiple informal teams that function to guide the overall development, to address conflicts in team efforts, to engage the larger organization in the effort, and to integrate formal organization with the informal design process. These teams form the backbone of an organization's design effort.

- **Advance team.** As the issues and solutions are investigated, it is important to have a group that oversees the basic direction, the impact on the overall organization, and the resources dedicated to the design process. Based on John Kotter's introduction of an informal *guiding collective* to accelerate solutions within organizations[40], an advance team is a mix of management and lower-level employees into a single team that keeps an eye on the process. The management members can be drawn from different departments and oversee resources, whereas other team members have knowledge of the work or business environment involved.

- **Design teams.** The actual members involved in the details of the design process form the central set of design teams. There may be more than one if the domains do not overlap: If one domain focuses on accounting solutions while another focuses on customer service, these can be divided into two design teams. Both may be monitored by a central advance team.

- **Users.** Active participation of current and future users is integral to the collaborative design and development of solutions. Research has shown that direct participation increases user satisfaction of the results, and when functionality meets user expectations.[41] With members of the community as part of the hub and ongoing interactions with the larger user base, much of the design process can have direct participation of the people who count the most.

- **Panels.** A number of panels can also be formed to focus on keeping the organization informed of progress, to heighten awareness and priority of the project, and to administer awards and recognition to the members involved. Each panel is composed of members who are closest to these activities. For example, as teams excel in their design undertaking, an Awards Panel can reward their work.

These different components of the design hubs provide ways for the formal organization to oversee the use of resources on the projects. Panels can create feedback to managers as bonuses, promotions, or raises are being considered. In this way, the formal hierarchy can tap the benefits of the creative power of the informal organization while ensuring that the project is overseen properly and that teams are acknowledged and rewarded for their work.

Composition

Although the project itself may originate at different levels or locations, the people involved in the design and development process are the ones located the closest to the work at hand. If it is a customer-oriented system, then it is those working closest with the customer. This begins to form

the hub. As discussed in *Types of hubs* on page 49, there are many different types of hubs, but for the design process, the first step in organizing a hub is identifying its members.

- **Users**. Where the system is aimed at a specific part of the organization, those members of the workflow closest to work should be identified. Where the system spans different locations, members of the staff at other locations should also be identified.

- **Customer or prospects.** Where customers or prospective customers are impacted by the proposed system, a key group of current customers should be identified. They may have the most impact on the organization's success and typically have been clients for the longest time. They may be brought on in a consulting role.

- **Information technology.** At a certain point, IT will be involved in the design and development, especially when systems need expertise in security, compatibility, and strategic direction.

- **Advance team**. Where a section or department is seen as sponsors of the system, members of the management team should be part of the advance team.

- **Facilitators**. Since the design process involves collaborative team interactions and new group dynamics, it is helpful to have facilitators who can guide the processes along. They can be outside consultants, or members of the participating departments.

Not all members that are identified are involved in all aspects of the design and development process. There may be a core group that handles most of the basics, and a larger collective reviews their work. Facilitators can work with the members to include other members in the process.

In a dispersed organization, especially one that has members of its workforce working remotely, the design teams need to be connected. Enter the *digital design hub*, which uses LCNC communications platforms to structure their projects and integrate collaborative design into the actual mechanics of their interactions.

Digital design hubs

Once the basic hubs have been identified and formed, the next step is connecting them to an online communications platform. These can be e-mail and video conferencing, but the most effective platforms centralize discussion threads, direct messaging, conferencing, and custom processes onto one platform. In this way, hub members can have an efficient online space for all of their various forms of communication. These become the organization's *digital design hubs*.

There are many reasons to have everyone connected in this way:

- **Dispersed workforce.** When members are working in different locations or from home, the digital platform provides a convenient way to discuss designs across time zones and locations. These hubs meet online and can continue their collaboration offline with other digital toolsets that keep the momentum.

- **Accelerated timeframe.** With the ability of members to connect with one another at any time and from any location, the design process is accelerated. Topics can be introduced, refined, and brainstormed from the comfort of one's home. Prototypes can be viewed and comments provided by a larger user base as needed.

- **Increased acceptance.** With a wider user community involved, issues and suggestions can be identified and addressed earlier in the process. This means that the final product has a greater degree of being accepted and implemented once it enters beta testing.

- **Digital repository.** The digital platform provides a central location for discussions, documents, history, membership, and information about various projects. These can be connected with the different phases of the collaborative design process. They can also contain access to prototypes, test results, and reviews.

- **Real-time updates.** News about the project can be easily distributed over the platforms so that groups can keep up with the progress being made. If meeting dates change, members can be updated and reminded through notifications.

- **Historical record.** The digital hub provides a record of conversations, ideas, and mandates that are useful in charting progress and understanding the evolution of a solution. Should the conditions change, and other factors enter into the decision-making process, the history can inform the next steps.

- **Oversight.** With all communications and documents in one location, a *guiding coalition* can oversee progress on any given project. In most cases, conversations are transparent and shared with members of the design hubs which include the guiding coalition.

- **Employee/team performance.** Because of the digital records, the various panels have access to how members are performing. This helps bolster an employee's prospects for a raise or promotion, as conversations, work contributions, and team efforts are in clear evidence on the platform.

The digital communications platform does not exclude face-to-face meetings, phone calls, and general e-mail. These are still an important aspect of the design process.

Digital collaboration

In addition to having a central platform for communications, digital design hubs enable the teams to implement the collaborative design process. This means that teams can identify, refine, brainstorm, select, plan, build, and test in measurable, succinct phases online.

For example, to identify the top targets of the design, the team can be assembled around an open-ended question that allows for a greater range of possibilities. Screens may look something like this:

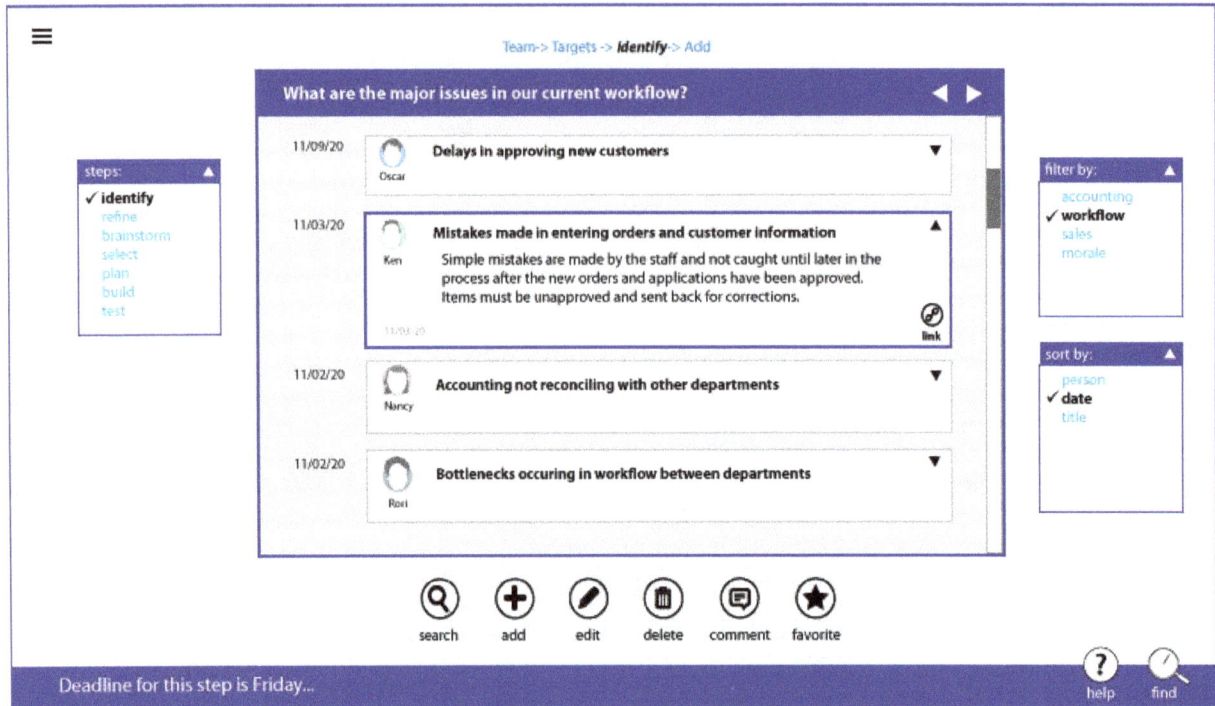

Each member can contribute potential targets of the design, and the team can refine each before settling on the top candidates.

When the main issues are identified, the team can further refine the details that help hone in on the components that need attention. In the later step of brainstorming, the team can suggest the details of the solution. From this list, they can upvote the individual features with the most priority. From this list, solutions to the original targets are selected.

These selections can be shared with other members of the enterprise so they see progress. This is the wider hub beyond the digital one.

Unlike a typical set of threads in a discussion forum, the design process has a sequential order, with results of one step linking into the inputs for another. In many cases, the steps are iterated until the final outcome is accepted. Facilitators keep the teams on track and can organize schedules.

Final thoughts

Design hubs form the backbone of an organization's design effort, and can include internal stakeholders from multiple teams as well as external users and customers. Digital design hubs are a centralized solution to ensuring an efficient and collaborative work flow.

15. Does Anyone Know How Things Work Around Here?

No one person really knows how things work in an organization. Upper management knows the overall results, the impact on business, and its customer base to some extent, but they have little understanding of the intricate details that go into taking orders or servicing clientele.

Members of the hubs know the work and the steps, but not always the personalities of the customers, unless they come into daily contact with them. The marketing arm of a company sees the needs of their customers and the potential in the marketplace, as well as how their products and services will fit.

Each hub is centered on a few key processes, with members contributing their knowledge and skills. It is only when everyone puts their heads together that a full and complete picture of how things work comes together. And with collaboration, these individual steps can be made more efficient and the group more cohesive.

The approach to energizing the hub with LCNC technology can do wonders. However this does mean taking on a larger project in the midst of carrying on day-to-day operations. Expectations must be tempered when it comes to assuming things can continue on while new software is being developing by the same people being relied on to do the work. But the simple act of including users in the design and coding can also energize the members; they feel like they actually influence the end product so it genuinely works and ultimately reduces the workload.

Warning Signs

Just the notion of introducing a faster and more sustainable way to do things can energize the various groups in an organization. The process can informally begin by considering the end users themselves. There are often several reasons and warning signs to consider.

- **Decline in quality.** By far one of the most important reasons is quality of the final product or service that an organization produces. This may be evidenced by returned products, poor ratings on customer service surveys, or decreased sales. However it reveals itself, the fact that quality is slipping can trigger ways to improve the process. Implementing a software solution might help.

- **Cost overruns.** If projects run over their budgets or the cost of producing products begins to increase, it may indicate inefficiencies in the process. There may be bottlenecks or additional staff called in that were not originally needed.

- **Increased volumes**. Due to the market or types of offerings, volumes for services and products may continue to rise and cause bottlenecks or delays. This can be a sign of anticipated growth by the organization. It can also signal increased calls for assistance.

- **Staff turnover**. An organization may see an increase in staff turnovers that may reflect issues encountered in the delivery of services. This can be that the staff is overwhelmed, or must work too many additional hours, or that they receive little support in areas that are hard to navigate.

- **Staff overtime.** In the same way, the number of hours that staff must work overtime, even seasonally, can indicate overloading. It is a sign that the actual process could use some help to reduce the staffing.

- **Increased competition.** The organization must see outside competition on the horizon. Usually new startups have the funding to invest in automation, so this can be a warning.

- **Increased complexity**. The complexity of a product or service can also impact its delivery.

If there are numerous options, matching the right ones can be time-consuming and result in mismatches. Automating the fit between options can reduce the staffing required for this process.

These can all be warning signs that spending some time and effort to evaluate additional software might be advisable.

It is always hard to know just how and when to officially begin a design process. So instead, another option is to begin to explore the possibility informally. This can happen in staff meetings, where management may be curious about what can be done to help out. Or, hubs can talk amongst themselves and offer suggestions for next steps. Either way, an informal design process, where the needs of the organization are addressed, can be started by members.

Using hubs to design and code

Organizations experience a number of pressure points as they relate to information flow and analysis:

- Information is not readily available, in the right format, or easily shared.
- Decisions made by the teams can conflict with the formal organization.
- Information many times changes or is not updated consistently.
- Organizations that span multiple sites may run into localization issues.
- Exceptions eat into the efficiency of technology solutions.
- Manual manipulation of data is still necessary and impacts quality and efficacy.

One way of overcoming these obstacles is with the use of LCNC technology. It does so by empowering hubs to develop technology using Agile methodologies. Information can be assembled, maintained, and presented to teams in ways to engage them to their work. This approach supports collaborative design, decision making, the open exchange of information, and the ability to communicate in social contexts with user-generated content.[42] In doing so, it allows staff across different specializations to interact with one other.[43]

How to be agile

Agile software development has evolved from its formation back in February 2001 when seventeen software practitioners assembled in the Wasatch Mountains of Utah to put together the Agile Manifesto. Up until that point, it was an informal approach practiced across industries to accelerate the development process and get software in the hands of users more efficiently. This also meant ensuring that the code was as bug free as possible (though there will always be little bugs creeping around), and that the features resonated with its users.

Agile practices and approach emphasize jumping into the actual coding as soon as possible by assembling the major stakeholders together to actively design and code. Traditionally, requirement documents are assembled by technical writers and their editors. This involves interviewing an assortment of users individually, writing up formal requirements documents, and a review process that takes a time and effort.

Requirement documents usually come in two flavors: functional requirements and technical requirements. One follows the other. Once the functions are outlined, the IT staff weighs in with the technical requirements of how the new software fits the security of the organization, its interfaces to current automation in place, and whether it fits into the technical landscape enforced by the CTO or CIO. So if the functional requirements of the users change—as they often do—the technical requirements must be reviewed and changed as necessary. These two forms of documentation must then be signed off on by the organization.

These requirements documents are then translated into a formal design document which can articulate database designs, technical layers, data interfaces to other systems, and security overlays. Where the functional requirements can be read and understood by the stakeholders, the technical design cannot. As it is aimed at the IT programming staff, a detailed understanding of the business and operational side of the project is not required to implement the technology.

The process of producing these requirement documents requires a large investment of time in both detailing the requirements and in producing an articulated design housed in the project's documentation. There is no guarantee that the documents will address the needs of the customers (since normally they are not involved) or of other departments.[44]

The agile approach minimizes these documents and places more emphasis on the prototyping process where versions of the final software are constructed quickly by the actual users. This can mean that the customer end of the software involves top customers in the design aspect. Agile development supports the technical aspects of the final software, but without the heavy reams of documentation. It relies more heavily on schematics, tables, and outlines. This shortened form is usually easy to assemble and maintain.

In this way, the software development processes start coding quickly, allowing it to understand the developed product. With LCNC technology, users can drag and drop elements of the product onto their screens to demonstrate potential design. Instead of lengthy documentation, which may be obsolete by the time it is rolled into a development phase, an agile approach is on much safer ground.

Finally, when customers are involved, their emphasis on easy-to-use, bug-free software becomes front and center. This means that not only are features designed but the testing of software begins. With testing part of the development process for each iteration, many of the quality concerns can be fleshed out and corrected early on. In traditional approaches, testing (whether it is alpha or beta testing) happens at the end.

This approach has also been used in other software development methodologies including scrum, lean product development, extreme programming (XP), DSDM, and adaptive software development, to name a few. The foundation of each of these is its use of an agile approach to the design and implementation of software.

Low-code/no-code development is a perfect fit with the agile approach. LCNC technology can link teams in multiple directions (vertically and horizontally) in the organization as well as across physical locations. It has been used in small and large organizations (e.g., transnational corporations) as a means to expedite decision making and bring groups together.[45]

Such systems must incorporate the ability for teams from different localities and organizations to jointly design their system.[46] They must also establish a more informal set of communications channels.[47] Using existing e-mail and formal channels can undermine collaboration. The focus of the systems, sometimes referred to as Computer Supported Cooperative Work (CSCW) applications, is not on automation per se, but rather on the support of social activities and methodology used by teams.

LCNC software aimed at smaller hubs generally does not replace existing systems in use throughout the enterprise. Rather, it serves to: link disparate systems together by providing information overlays; provide informal communication channels where teams can discuss issues privately; provide an easy-to-use view of progress and outstanding items; and keep a thumb on the pulse of the organization. Each hub can have its own customized center. Members can invite others to the hub. Teams focus on the most critical topics.

LCNC software is also mobile. This means that information is secured but allowed to be accessed at any time from a phone or tablet, giving team members around-the-clock access and encouraging hubs to meet at any time.

Final thoughts

Bringing a new product or service into existence means there are a lot of hands involved, and participants don't always know what the others are doing at any given moment. There are several approaches to the workflow that can be useful, with certain hub members being designated to understand and execute key steps. Hubs may need to consider end users in the design process and how current products are being used and perceived. An agile approach that can work in tandem with LCNC programming is an effective combination for making sure members not only get involved but stay involved in the design process.

16. All Aboard!

It's time to start a sample project.

Traditionally, formal projects are approved by management, who in turn allocate the necessary resources, set goals, and approve schedules. A project team may be assembled with a lead who reports directly to management. Status reports are generated and progress is tracked against the formal schedule.

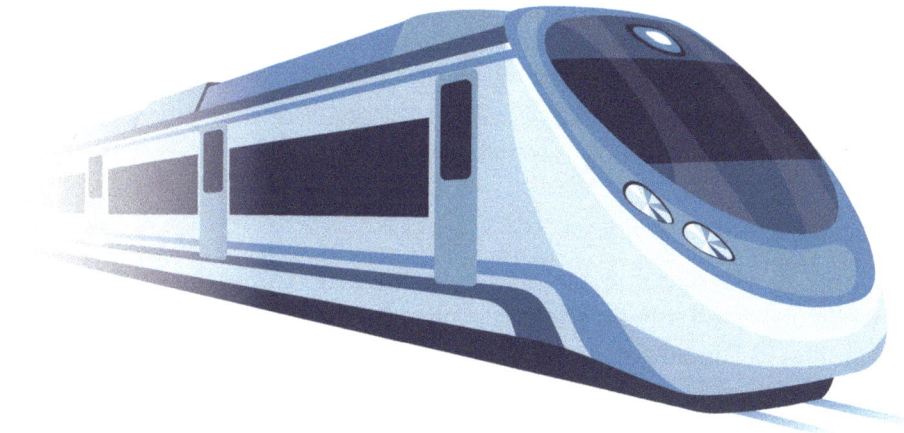

This approach does not always produce the desired end products. With limited time and without the key players involved in the process, the final application may fail to solve the issues or lack the features needed to be effective.

By contrast, designing applications often informally begins with issues or opportunities that catch the attention of someone on staff. Perhaps a customer may make a suggestion that takes hold. Instead of waiting until the idea finds its way to the upper echelon for approval, work can actually begin earlier with informal hubs getting involved. This can happen as a result of a comment made at a meeting where an issue or opportunity is mentioned. The manager in charge may express interest in learning more and ask members of his team to look into it. This in itself can get things rolling. At this stage, resources have not been allocated, but it is one of many ways designing starts.

By starting an informal project earlier with informal buy-in from management, in the end it may take a shorter amount of time to reach a decision and use fewer resources. In effect, the design hubs research the issues and brainstorm solutions even before a formal project is authorized. Based on their initial findings, more resources can be allocated.

As informal parts of the organization with management support, design hubs cross departmental lines. Thus, starting up a design process to see if it makes sense to take next steps can involve a cross section of the workforce. Since it is not mandated or assigned as work, members can prioritize issues as they see fit. It is helpful to outline the types of information that is handy to have, and encourage members to discuss the prospects with others in their hub.

Each member has a unique perspective on the day-to-day flow of work. Traditional design work begins with IT and requires management to approach the project and budget. With design initiated by the hub members, the process can begin as a way to understand what is not working and how to solve it, or to see if a potential opportunity is worth exploring.

Advance team

The first steps in organizing a design hub are identifying its central members. To this end, it is important to have a well-connected advance team in place (usually from the management team of the involved departments). [48] The advance team acts as a bridge between the informal design hubs being formed and key managers in the formal organization.

Their tasks are to begin discussions of how get the design process started, and who in the overall organization would know the most about the issue or opportunity at hand. The assembling of the design hub can be handled by those closest to the work. It is important to focus on the targets, and initial discussions of the advance team should focus on the following topics:

- **Mandate.** If the design process has been initiated by one or more members of the management team, then the mandate of the advance team should be clarified for its members. It's important to know the general arena of what the focus should be.

- **Initial targets.** With a domain understood, the team should phrase their targets as questions. Targets form the basis for investigation. For example, if there are problems in signing new customers up with the application process, then the team may phrase it as, "What types of problems arise in signing up new customers?"

- **Reaching out.** Finally, the team should reach out to members of their organization closest to the work at hand for them to actually assemble the design hub. In other words, instead of the advance team appointing members to a task force, it is kept more informal by asking these members to bring in people they feel would be helpful.

- **Setting expectations.** The goal is to design a piece of software that might correct a problem or take advantage of an opportunity. It starts with analysis done by hub members. But the members of the evolving design hub must also understand that this can lead to bigger things.

- **Facilitators.** It is helpful to include facilitators in the process as it starts. The role of facilitators is to guide the design hub through a collaborative process. This will help keep the design hubs focused and ensure that the collaboration is open to a wider set of perspectives, suggestions, and involvement.

- **Venue**. It is helpful for the advance team to meet off premises where their discussions can be more informal and not follow a preset agenda or structure. Rather, it begins as discussions and ideas of how to build the design hubs to address the issue or opportunity at hand.

The advance team lays the groundwork for next steps. Although they do not directly appoint individuals to the design hub, they approach those that can. These individuals have an understanding of the work and are well connected to others that do.

Shared folder

Early in the process, the advance team should begin using a simple shared folder to facilitate ramping up the design hub. Inside the folder, the advance team can organize key documents such as spreadsheets, text documents, notes, and slides as needed.

These shared spaces are generally available through major technologies provided by Microsoft, Google, Dropbox, and others. For most organizations, shared folders are already in use. Setting aside a separate project area or folder allows the advance team (and the design hub itself) to stay organized, collaborate on lists and documents, and understand the progress being made. These documents may include:

- **Spreadsheets**. A favorite tool among members of organizations is the *spreadsheet*. It can provide a shared list of steps, fields, priorities, and generated documents. When created and shared in the design process, they provide a collaborative vehicle that accelerates the process and engages a larger portion of the final user set.

- **Shared documents**. As descriptions or details are collected, it is useful to have documents that can be viewed and edited by the advance team and the design team.

- **Slides**. Shared slides are a quick way to present information at design meetings as well as summarize the targets and solutions in an easy-to-understand format.

- **Meetings**. Real-time conference tools are extremely helpful in organizing meetings with team members or with the end users.

- **Surveys**. Surveys are quite common these days and can add to the understanding of the presenting issues, possible solutions, and basic views of the user community.

Many LCNC providers have these tools embedded in their platforms. If the LCNC platform is still be considered, there are well-known toolsets from Apple, Microsoft, Google, and others.

When these documents are shared in a secured, cloud environment (or on a common shared drive), individual members can see the current status. Depending on the setup, team members and end users can also contribute directly to the folder.

Assembling the hubs

Members of the advance team began reaching out to those that might be interested in pursuing the design of a solution to the matters at hand. Individuals may be called in to meetings, or discuss it over a casual cup of coffee. But in the end, a group of people are brought together with the mandate to work collaboratively toward a set of solutions to the issues at hand. The basic steps in forming the design hub are the following:

1. **Identify key people.** The advance team should build its initial list from those in the organization with the best knowledge of the work being addressed. These can be people who have reported issues before or who have submitted suggestions on how to approach a solution.

2. **Talk with supervisors.** Supervisors may also have people that they know in their area with expertise in the issues being addressed. They can also provide a list to the advance team.

3. **Talk with customers or third parties.** Many times it is effective to include customers or other third parties outside the organization for their opinions. The topic can be one of interest that is mentioned to customers who may have some suggestions. Whether they are part of the design hub itself will rest with the design team as they are assembled.

4. **Use casual meetings.** Members of the advance team may also choose to meet casually with staff that seem to have an interest in—and a good understanding of—the issues. For example, members of the advance team can individually meet with a customer support co-worker to get their perspective. It can begin to shape how the advance team begins to facilitate the formation of the design team.

At the end of this phase, members of the advance team have lists of potential design team members. At the start, the core team of designers can be approached as an informal project with a set of targets to explore. If there is interest, these members can build out their design team and include others from inside or, with the advance team's approval, from outside the organization.

In order to jump-start the process, the advance team gives general guidance in pursuing the project. In most cases, since the advance team has the approval of various departments, the design team can arrange times to meet that do not interfere with their work.

Going digital

Once the core members of the design hub are in place, they reach out to others in their informal group to join up. This is principally done through hub meetings which are typically conducted online. The first step is lining up the digital platform to use.

- **Online conferencing**. Real-time video conferencing has become the norm in how hubs meet. There are a number of packages that are available. Before deciding, members should see if their organization has a platform for this and if it provides the features that are needed.

- **Digital forums**. A digital forum is a discussion that can be organized by groups, topics, and threads. For example, public forums are many times conducted using the Wordpress platform and a number of plug-ins. For organizations, similar features are available.

- **Shared folders**. Folders should be provided for shared materials. These can be different than the advance team folders, depending on how the hub wants to proceed.

- **Notifications**. Real-time notifications should be part of the digital array of features. This allows members of the hub to be notified as needed.

- **Privacy**. Both the conferencing and forums should have strong controls to regulate access. The core group should decide on administration, roles, and access. As the hub expands, more members are brought onto the digital platform, so it needs to be relatively easy to add new members.

- **Security**. Security should be set up to prevent outside exploitation or attempts to hack into the platform, since subjects can be sensitive and involve proprietary information.

There are a number of LCNC platforms that support these features. Often, if a LCNC development platform has not been selected, the hubs will use facilities already in place at the organization.

Once in place, the design hub will make its transition into being a digital design hub, since a large portion of their interactions occur online.

Example of a LCNC Development

Let's use the following example from a hypothetical advance team given the mandate to explore solutions:

> The process of applying for new financing for current and new customers is slow, causing many to go elsewhere.

This is an issue of workflow and is typical of organizations. It may start in the financial organization with the calling officers, the staff who are typically responsible for bringing in new customers. But it quickly is engaged in a workflow with the staff that process the requests, then evaluate the credit worthiness and security, then produce documents for review and signatures, and finally to staff that can underwrite loans.

Pulling these members into initial meetings may seem straightforward. In many ways, the organization already knows who is involved. There may also be opinions as to who is not pulling their weight in the process, or how managers can avoid slowing down the process. But these perspectives take a backseat in organizing the design hubs.

The same is true of lists of members in the hub and their roles, as shown below. Using a shared spreadsheet, members of the advance team can add their suggested participants.

People & Roles (Example)

Person	Role	Comments
Robert Smith	Loan originator, Loan approver	Bob serves a couple of functions
Janet Doe	Loan originator, Receptionist	Janet takes basic information from applications and hands off to others.
Pat Jones	Loan approver Property assessor	Pat approves loans for next steps and can establish valuation on properties
Lucy Adams	Credit officer	Lucy handles customer credit rating and lines of credit
Mark Williams	Technical support	Mark supports desktop hardware and software for the department

Person	Role	Comments
Janet Brown	Loan supervisor	Janet oversees loan process, handles exceptions
Mary Miller	Compliance officer	Mary makes sure the process adheres to regulations and compliance

The list is a good first step to building the design hub. With larger hubs, it can be less efficient to involve all the members in the design. Everyone may have their own way of doing things, with separate recommendations on how best to integrate items into the technology and with their own sense of priorities. Yet breaking them into smaller teams with specific design steps may facilitate the engagement of a large portion of the organization's informal networks.

There may be more than one hub involved in the work, so this may break out as people and roles are added. The advance team may find that suggested participants span different locations. The problems encountered in one locality may not be present in another.

The benefits of the design process are still quite clear: an initial review of the issues has produced some actionable steps the organization can take.

Joining Up and Connecting Up

It is still an informal project, since management has not officially allocated resources and there are no deadlines or even talk of a final software solution. Still, the advance team is interested in giving the design hub space to explore the topics at hand.

One of the easiest ways to get going is through the use of online group communication platforms. If there are existing platforms, such as Slack or Teams, in place, then the initial discussions can happen online. If not, face-to-face meetings over platforms such as Zoom, Meet and Skype can begin. Some guidelines for the first meeting of the advance team include:

- **The launch**. The first meetings are introduced by a member of the advance team outlining the interest they have in a particular set of issues or opportunities.

- **Independence**. It should be stressed that the design hub is independent of the advance team, and can pursue the topics in their own way. However, it is recommended that the process be collaborative.

- **Framework**. If there are facilitators available to assist, they can be introduced and outline the basic framework to be used. They also describe their role in the hubs.

It is helpful to have an agenda for the first meeting. This can simply be to introduce the topic and coworkers to one another. For our example, a simple agenda is used.

Agenda (Example)

- Exploring the issues
- How can we improve our service?
- First steps
- Informal team
- Your questions

Once the initial meeting sets up the expectations and framework, the group can decide when and how to meet next. These discussions can cover a range of topics, including:

1. **Defining the hub**. From the list of suggested people, the hub can meet to review the issues and expectations as outlined by the advance team. The main focus is defining a hub that can address the different facets of the problems. Initial members may add more people as issues extend beyond their immediate expertise. These additional members should be invited to participate as time permits.

2. **Venue and times**. Once the hub members are fairly well defined, venues and times should be set up for meetings. These can be during office hours, or at other times as the advance team advises. Folders should be set up for design documents and spreadsheets to help organize the hub as it begins its work.

3. **Rules/Guidelines**. A set of rules and guidelines should be included in the first meetings. Although this seems like too much structure, it will come in handy when there is a disagreement or conflict. It is natural to have disagreements, but you want members to respect each other's opinions, to stay on topic, and to give equal time to members.

4. **Topics**. With the teams in mind, the next step is formulating an agenda. Here, the facilitators can get an initial set of topics and route these to the members. The members can add or combine the items. If the agenda items are too detailed, then the main topics can be used to open up discussions.

5. **Targets**. The main goal is a list of targets. These can be issues or problems encountered by the group that can be the subject of further investigation down the road.

6. **Drilling down**. The list should provide examples and details, situations where problems or obstacles have existed. The target may also be a list of areas that would greatly benefit from the introduction of a more automated system.

7. **Time limits**. Set goals for how long the discussions should last.

8. **Rotating leads**. The leadership of the hub should be able to rotate to different member based on the topics being discussed. For example, if accounting compliance is a topic, the lead should shift to members with the most knowledge of the subject. This rotating lead is sometimes referred to *collaborative leadership*.

These initial steps help the design hub to gain its focus and set a pattern of how meetings will be run.

Involving the stakeholders

The development of a solution is not isolated to the design hub. With larger organizations, users and stakeholders must also be involved in the proceedings. To identify the stakeholders, the advance team should plot out the departments and parts of the global organization where the application will be used, and where the results of the application will be sent. These individuals form the network of extended stakeholders. There will be points during the development and prototyping that the designers will want to include the stakeholders. The manner in which the stakeholders are involved is left largely up to members of the design hub. There are a number of common ways to include the stakeholders:

- **E-mail updates**. In the early stages, simple updates sent out by the design hub to the wider audience will alert them to the process underway. The composition of the e-mail and frequency can be determined by the advance team.

- **Videos**. The leads on the project can also put together short videos describing the progress on the project. If designers have questions or need assistance, this is a useful way to reach out to the audience.

- **Slide presentations**. Slide presentations can also drill down into the details of the process. It is helpful to reiterate the goals so the stakeholders understand the domain being tackled.

- **Live meetings**. With video conferencing, design teams can make presentations to a wider audience, showing prototypes as they are ready or outlining the main features. They may also use this method early on to involve the stakeholders in the process of problem solving and brainstorming.

- **Group messaging.** Using the LCNC communications platforms, smaller group meetings that involve stakeholders can also be held. Discussions can hit on the critical aspects of the project.

- **Design and coding.** With the advent of LCNC development, many stakeholders are in themselves designers and coders (sometimes referred to as citizen coders). For applications that do not require low-level programming languages, stakeholders can participate in the process themselves. This is particularly important during the prototyping phase.

Stakeholders should be included in the process and kept informed along the way.

Final thoughts

The beginning of a new project is always an exciting time. Bringing together a team, identifying an issue that needs a solution, and brainstorming ideas can all begin informally with the right people involved. It's important to follow a number of steps even before a formal project or pitch can get underway, and involving key stakeholders from the start ensures projects deliver on time and with all the necessary input from relevant members.

140

Part VI: Refining Targets

141

17. Taking Aim

From our example, the digital design hub has been assembled. It is time to identify the specific targets of the project. Not all areas outlined by the advance team can be addressed, and many are too general to develop an action plan: the first step is to begin detailing the targets.

Targets

Targets are examples of the issues or opportunities that fall under the mandate of the project. Teams meet digitally to begin discussions on just how the issues manifest themselves in the daily work. The design hub can reach out to the wider organization to solicit input, but it is the members that organize these targets.

In our example, the advance team has given the basic mandate:

> The process of applying for new financing for current and new customers is slow, causing many to go elsewhere.

From this, the process begins with identifying the target issues. The targets are assembled and listed for the members to view based on previous discussions. These can be detailed or clarified as needed.

Targets (Example)

Priority	Issue	Target	Comments
1	Long waits to get correct proposals to prospective customers	Increase the ability of the staff to handle larger number of proposals	Large number of declines with prospects at the end of the line
2	Mistakes made with quoting terms	Eliminate the common mistakes; reduce entry of data if possible	Wrong quotes are sent out, making it harder to correct and keep the prospect
3	Unable to generate documents to sign since data has mistakes	Clean up the data entry and credit checks	Documents have errors so clients must wait for corrections many times
4	Applications received are slow and filled with inaccuracies	Revise the applications	There are fields that aren't needed or can be misinterpreted

There may be a longer list, but the top targets are selected by the team based on their perceived priority. These priorities are established by the team members, typically through voting in a platform's forum. Issues may have more than one target, but wherever possible these targets are refined with the support of the specialists.

Refining

From the initial list of targets, the design hub begins refining the issues, with the goal of understanding the underlying causes of the issues. With hub members who are familiar with the different aspects of the process, a good first step is listing the tasks involved. This can be done using a shared spreadsheet, with hub members contributing or asking questions. Where the step is performed by another system, it can be indicated on the spreadsheet. Where there are specific inputs involved, these can be summarized and detailed later when the design shifts into actual coding. For our example, the final task list that can be used to develop the solution might look like the following:

Task List (Example)

Seq #	Step	Who	Comments
1.0	Receive inquiry	R Smith, J Doe, P Jones, S Lewis	Application for loan arrives by e-mail or phone, and is transferred to the loan team
1.1	Interview applicant	R Smith, P Jones	The team member discusses needs with the staff
1.2	New contact created	J Doe, P Jones	A new contact is created and stored
1.3	New application started	R Smith, J Doe, P Jones	An application is started with links to contact, summary of loan needs, type of loan, desired date, and other information

When this is information is in a shared spreadsheet, anyone in the hub can review it and contribute. It is also important to reach out to stakeholders for their input and comments. With larger organizations, the participants of the workflow can be quite large, so identifying them by name is not recommended. Instead the formal work groups can be shown.

Documents

When working with workflows, it's good to identify the types of documented need upfront. For example, when an applicant for a loan contacts a financial company, the staff will first want to send out a general welcome letter and outline the next steps. If it is a recurring customer, some documents (e.g., the profile) are probably not needed. This can be sorted out later, but having a list helps to identify the outcomes.

Document List (Example)

Document	Use	Comments
Welcome Letter	To reach out to prospective applicant	In response to a phone call or e-mail, it identifies the central products offered and welcomes the prospective applicant to take next steps
Loan Application	Initiates the review process for an applicant	Currently filled out by the staff but this can be filled out by the applicant online
Privacy Release	Accompanies the loan application to release company from liability	Must be signed by applicant (and co-applicant if applicable)
Applicant Profile	Filled out by applicant and co-applicant with phone, address, etc.	Accompanies application, used for contact

Examples of these items can be assembled in a separate folder. There may be new documents or reports that are requested or required; these also can be drafted.

It is also useful to identify the inputs, that is, how a particular set of tasks gets initiated.

Input List (Example)

Input	Purpose	Comments
Inquiry	(E-mail or phone) Usually to ask basic questions before any next steps	
Request for additional information	(E-mail or phone) Usually indicates that the prospective applicant sees a good fit with their needs	
Request for next steps	(E-mail or phone) Asks to speak to a loan originator	

The Input spreadsheet is also helpful to understand more about the current and anticipated work volumes. This can be discussed by the teams during the design process.

Sketching out the workflow

One of the more useful steps in designing the technology is charting the steps involved. This follows the tasks undertaken by the group and is presented usually as a simple flowchart. The resulting chart does not need to be fancy; it may be hand-drawn or produced by simple drawing applications. (Many LCNC platforms include the ability to construct workflows as part of their inputs.) The goal is to identify the major steps in the process, any documentation produced, and decisions that are made along the way. Members of the hub that are involved can collaborate to produce the flowchart, and it is intended to be easy to understand by the users.

Spreadsheets are an excellent way to tie everything together in an easy-to-understand diagram. The workflow does not need to be fancy. Simple icons help but are not required. Annotations are very helpful for the documents produced.

Consider the process of extending a loan using spreadsheets created by the team. There may be different parts of the enterprise involved in this initial step. In looking at the actual application however, a common workflow across departments can be outlined, one that ensures that the applicant qualifies and that the underlying security (e.g., the property) has a valuation that supports the loan.

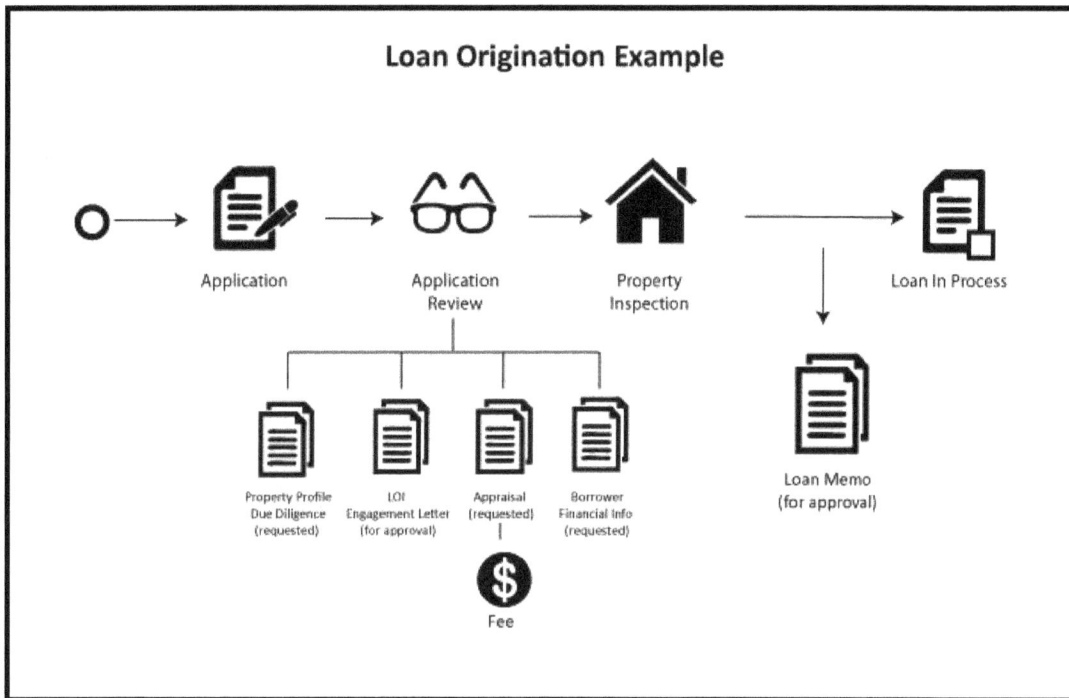

Loan Origination Example

Application → Application Review → Property Inspection → Loan In Process

Application Review:
- Property Profile Due Diligence (requested)
- LOI Engagement Letter (for approval)
- Appraisal (requested) — Fee
- Borrower Financial Info (requested)

Loan Memo (for approval)

Along the way it may become clear that there are exceptions: these can be noted or included depending on how often they occur in normal processing.

Notice that the individual steps can be quite general to start with. The *Application Review* step does not go into specifics, but from this review a number of documents for the applicant are produced. This relies on the judgment of the staff to produce the necessary letters and fees. At a glance, we can see the major tasks.

Another workflow chart can be assembled for taking next steps, such as once the loan is approved and needs to be funded and booked.

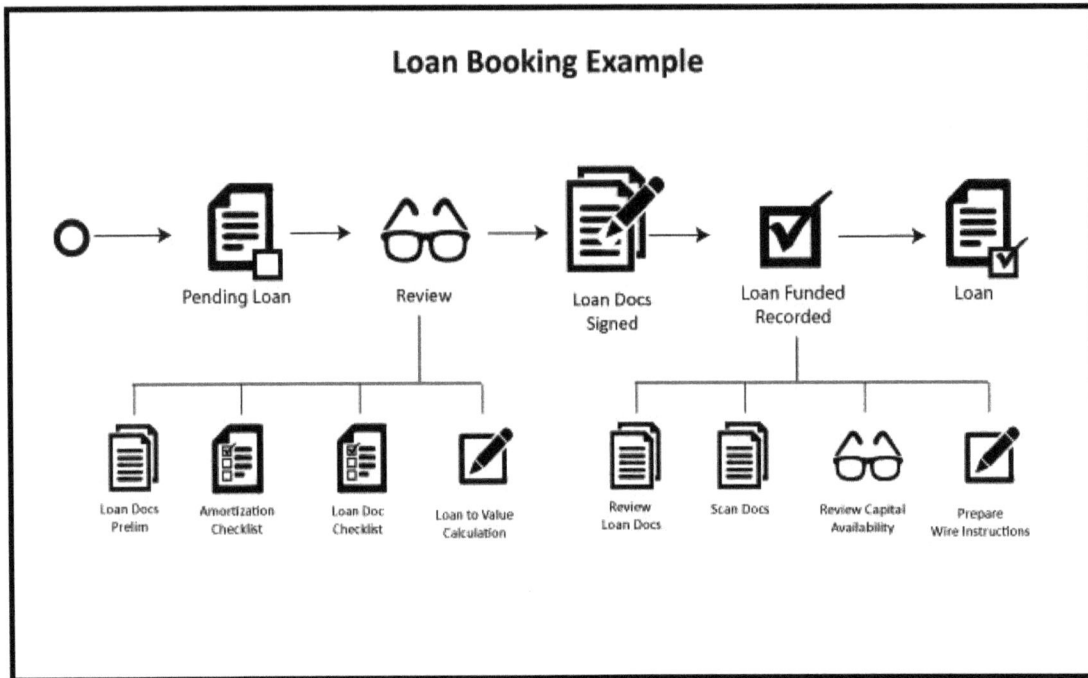

Loan Booking Example

At a certain point, the charts become detailed enough, and the process can be reviewed by the end users. Since workflow should be easily understood from the drawings, the members can confer in a meeting and make suggestions or corrections. The lists of tasks, staff, and documents serve to outline the work by providing a level of detail that only those who work closely together know.

Where are my reports?

This step in the design process is also a good time for middle and upper management to provide their reports on the list of documents and reports. They may want to keep the actual reports as examples in a separate folder. Management might be surprised to find out that their reports can be generated more easily and accurately when data is available in a central location.

Report List (Example)

Document	Use	Comments
Current prospects	To identify prospective customers who have called in	The list gives operations an idea of how well marketing is doing and helps project potential business
Current Applications	To identify loan applications submitted	
Current Loans	To keep track of total loans	Must be signed by applicant (and co-applicant if applicable)
Delinquent Loans	To insure that customers are contacted and assisted if their payments are delayed.	Accompanies application, used for contact

As with most of the lists, they can be revised and expanded as the design hub works through the refinement process.

Getting out the measuring tape

The refinement process is iterative. Having worked through and listed the different tasks, documents, and reports associated with the target, it is time to circle back. The question is to what degree the issues present themselves during the workflow. In many situations where there is a preponderance of errors, the target issues may be found in the documents and reports. On our example, where the delays are the result of errors, the errors should be identified in detail.

For larger organizations, it becomes important to measure the occurrence and impact of the issues on the organization and its customers. To this end, setting up measurements using various techniques can help the design hub to understand what is occurring. There are a number of ways of proceeding:

- **Initial surveys**. From the interviews, the hub can put together initial surveys to be sent to the different areas involved in the work. For example, the customer service staff has one set of questions, whereas the loan origination team has a different set.

- **Surveys**. The initial surveys are then refined by the hub members. These can be sent out by e-mail, or linked to notifications, to the staff and selected customers as needed. A number of products offer surveys through their forms, such as Microsoft and Google.

- **Staff interviews.** It is best to start with interviews with members of the organization, including the stakeholders. This helps to isolate the categories and zero in on the types of questions that the hub wants to pursue.

- **Customer interviews**. In many cases, the hub members may feel they need a more in-depth view of the problem through the customers' eyes. Although the target issues may seem minor, they may be considered significant by customers or prospective customers. Instead of working through customer service alone, the team can reach out to selected customers for their input.

- **Online meetings**. Interviews can be conducted via online conferences where teams are asked about subjects related to the issues. With the support of team members, issues can be clarified and examples given. Straw votes can also be taken to measure the occurrence or severity of the problems.

- **Review of organizational data**. Statistics on the issues can many times be found in the organization's data. For example, customer service may log a number of requests that specifically deal with the target issues. Whether the problem is resolved or should be referred on can be clarified. From this, reports of customer history can be assembled.

- **Exceptions**. As part of the process of measuring the issues, it is useful to understand exceptions. These are events that are not handled in the workflow or easily included in the documents and reports. Exceptions fall outside the typical processing of information.

By measuring the occurrence and impact of the target issues, the design hub can determine the likely causes of the issues. Where the issues are more sensitive or may be engrained in the organization's culture, surveys and questions may not deliver the objective measurement.[49]

Measurements serve a dual purpose: they can help determine the underlying causes of the issues, or help to understand how they present themselves to the organization. They can be used after the implementation of the software to determine just how well the solutions worked.

Action research has a long history of assisting with the diagnoses of organizational issues, dating back to research by Kurt Lewin.[50] It involves data collection as a means of collaboratively diagnosing problems. For the refinement phase, the data collection is used to understand and detail the targets. Later during the selection and planning phases, it can also be used to focus the solutions on the specific nature of the issues.

Final thoughts

Identifying issues to target in an existing workflow are a key initial step in the design process. Issues, potential fixes, and additional feedback can be collated in easily accessible spreadsheets, and housed on a digital platform for hub members. This also creates a seamless solution to sharing findings with stakeholders and other members of the company. Defining potential new workflows through charts and sketches work in tandem with spreadsheets to illustrate the hub's input and ideas.

18. Brainstorming: Get Your Umbrellas Out!

Brainstorming is a highly creative endeavor that requires that teams put on different hats, to see things in a completely different way, and to take chances with suggestions on how best to solve each target.

There are a number of ways to encourage out-of-the-box thinking, but one of the most effective can be settings and times that are unusual. Perhaps your team meets over breakfast, either virtually or at a diner with a large table. Many times we have our best ideas when we are fresh—so find a time when the hub is available, awake, and alert.

Since design teams can blossom out, it's best to break the teams into smaller "brainstorming subgroups" of five or six. These can be shuffled as the list of targets is attacked. These lists outline the basic issues that the software is aimed at eliminating or reducing. The design team uses them to focus attention on the priorities and the group's understanding of the causes of the problems.

Pick a venue

When the design team meets to brainstorm, it can be virtually (through Zoom, Teams or Meet video conferencing), in a conference rooms, or even outdoors. Choosing a space where participants are out of their normal environment is helpful to get the creative ideas flowing. After all, when focused on completing work, staff is not necessarily trying to find new ways to do things. They just want their contribution to be timely, without mistakes, and in a form that's useful to others.

But brainstorming is different. It looks for participants to throw out ideas that solve one of the target issues. It may be complete nonsense or it may be right on the money. Regardless, in a brainstorming session an idea should always accepted, explained where needed, and placed into a list of ideas for further consideration. With this as the goal, the venue should be relaxed, away from the usual grind, friendly, and comfortable.

Brainstorming teams

Ideally brainstorming teams are organized from the larger design team and should be broken out into smaller groups. They huddle together to work through the targets, throwing out ideas and discussing them.

The process should last 15 or 20 minutes. It's meant to be quick, and the emphasis is on the *quantity*, not necessarily the *quality* at this early point. Later as the whole design team begins to dissect and discuss the ideas on the table, they can be refined.

No idea is wrong

The general rule is: no idea is wrong. Even crazy impractical ideas are fine as they may trigger other ideas from members. A good rule of thumb is at least one idea from each member of the brainstorming subgroup; the more, the better. Each target is worked through fully by the team. One of the members should write down the idea to be discussed by the larger design team.

During this phase, team members should not criticize the ideas presented. They may however ask for clarification or examples of how a suggestion would be implemented. As much detail as possible is encouraged.

Sketching a solution

As part of the process, members may enjoy sketching their suggested solutions. There are a number of ways to do this. Simple paper and pencil work fine. If these are to be presented online, the member can take a photo on their phone and post it. There are also a number of sketch applications that are easy to use with a stylus on a mobile phone, such as Sony Sketch, Sketchbook, and Adobe Illustrator Draw. The final sketches can be exported as JPG or PNG files that are easily posted. Or, a member may cut and paste images into a word processing document, such as Microsoft Word, Google Docs, or other applications that permit images. Images can be labeled and exported as PDF files if needed.

The sketches do not need to be overly neat or complicated. The goal is to assist the team member in generating new ideas. Once the team member is satisfied, the sketch can communicate the idea visually to others on the team.

For example, an idea for trimming the time it takes to fill out an application can be illustrated with the templates that a prospective customer would use. This idea can be introduced as a sketch. Fields on the template can be mapped to those on the application, based on the type of application or the prospective customer. The idea is straightforward, but the sketch can show how only some fields map over to the new application.

Connecting offline

Although brainstorming using Zoom or other video conferencing tools is a great way to stir innovation, it can also be done offline and outside of the regular meeting time. A number of products provide the ability to discuss, refine, brainstorm, and upvote ideas using online forums or channels. Offline tools allow the design teams to link sketches, references to other documents, or research.

Drilling down

The brainstorming sessions can explore the details of the suggestions. This can be done through questions that the team asks each of the members presenting their suggestions:

- How would your suggestion work if not all of the information was available?
- How does it solve the underlying cause as you see it?
- Where does the idea fit into the overall software solution?
- Does it cover all of the instances that you'd expect to see?
- When do you see your solution being used in the workflow?
- Are there any reports that need to include the information your idea captures?
- Would users easily understand how to use your solution?

Depending on the idea, it may be more general and require more explanation to make it useable. Again, other team members should resist any critique of the ideas at this point. Instead, the goal is to flesh out the details. If there are any vulnerabilities, these can be discussed during the Selection process.

Ranking the ideas

Once the team has their list of ideas, it is time to begin refining their list. This list will in turn be presented to the larger design hub for further refinement. The exact number of ideas they take back to the larger group is flexible; they may see five ideas as having promise, though ideally they can reduce the list to three top candidates.

To rank the suggestions, the team should take the following steps:

- **Establish criteria**. How the ideas are judged should be based on identified criteria. The most important one is how well the idea would solve the underlying issue. Other criteria may be how easily it might be to implement, how flexible the solution would be, and how well understood it is.

- **Discuss each idea**. The team then discusses each idea or suggestion, asking for clarification or identifying potential vulnerabilities. The member who suggested the idea can modify their suggestion and defend its capabilities.

- **Ranking**. Each member then assigns points (e.g., 1-10) to each criterion for each idea. These are grouped by issue. Based on the number of points, the ideas can be ranked.

- **Top three**. The top three suggestions for each issue are then collected with brief explanations. The team can modify these for clarity.

- **Combining suggestions**. Many times ideas fit together under a single heading. For example, instead of suggesting that expiry dates on loans be calculated as well as monthly payments, these can be grouped, as loan parameters are calculated automatically.

Each team then has their list of suggestions for each issue in the target list. It is time for the design hub to meet to discuss further.

Discussing in the larger design team

Once the smaller brainstorming teams have completed their lists, the full design hub is brought back together to discuss the ideas. For each brainstorming team, one member presents the ideas and explains them further. The design hub can ask questions, and the ideas are added to a larger list. Each brainstorming team has its turn. Then the design hub begins drilling down further into what is entailed in the suggested solutions.

Ranking suggestions

From this list of ideas, the team ranks the ideas and lists them in order for each issue. Generally, it is the top three ideas that are brought back to the larger group for discussion and further ranking. The same discussion of criteria and rating of suggestions happens in the larger design hub, as described in the section above:

- **Establish criteria**. How the ideas are judged based on criteria, such as how easily it might be implemented, and how well it would work.

- **Assign points**. Each member of the hub then assigns points (e.g., 1-10) to each criteria for each idea. These are grouped by issue. Based on the number of points, the ideas are ranked.

- **Top three**. The top three suggestions for each issue are then collected with brief explanations, in much the same way it was done in the smaller teams.

- **Combining suggestions**. Many times ideas fit together under a single heading.

The goal is to narrow the suggestions into a manageable number. Care must be taken not to eliminate suggestions because they are considered too far afield.

Discussing the top ideas

From this list, the design hub selects the top three ideas for the target issue. These are the proposed solutions. The criteria for ranking these ideas can vary, but usually include how practical the solution is, how straightforward it would be to implement, if the organization has the expertise, and whether the amount of time involved is acceptable. Resources can be another issue, but generally this is left to the formal organization and to management once the list of solutions is presented.

Finalizing the solution

With its prioritized list of recommended solutions, the design hub begins to formalize the design solution. This solution incorporates the ideas that the hub have agreed on. Any questions or details that are available can be investigated offline, or left to the prototyping phase to try out.

Final thoughts

Brainstorming can be one of the most creative and engaging steps in the design process. Hub members break out into smaller teams to consider solutions to the identified issues, first individually then as a team. No idea is wrong, though it is still important to adhere to a structured discussion between members, to make sure all voices are counted and heard.

19. Selecting the Best LCNC Platform

With the lists of documents, reports, and workflows, and having seen a few off-the-shelf packages, the organization is in a good position to decide whether to proceed with a LCNC solution.

Management plays a vital role in asking questions about the net impact of new LCNC software on the performance and efficiency of the organization. After all, they will be the ones approving a budget and overseeing progress. To this end, management now has to look closely at the workflows and ask the right questions:

- Are there any duplications of effort by the staff in the work? For example, is customer information entered multiple times by different people, maybe in other hubs?

- Are there mistakes or errors that can be averted with a LCNC solution?

- Do we decrease the delivery time of our products with a more efficient approach?

- Should our teams include or inform others in the process as a way to further our results?

- Does a software solution improve the work environment, perhaps taking pressure off of staff or reducing the overload at peak times?

- How long will it take to put a solution in place?
- What are the projected costs?

These and other questions can be first raised for discussion. Ideally a larger portion of the organization can be assembled in an informal setting for their opinion. Management may not want to move forward or may want to tackle a subset of the tasks, but it is always good to reach out to the organization. After all, a number of the spreadsheets were compiled by the staff.

One way is to find a way to meet informally, say over lunch or during extended breaks. Hubs can be divided up and management can make rounds to ask simple questions. For example, an online get-together can be scheduled over lunch. It is important to keep the setting relaxed and for management to learn more about the day-to-day efforts from the staff that actually performs the work. By asking questions with drawings or drafts in hand, management can learn more about what makes sense as next steps.

What's the price tag?

If management is going to approach a new software project, even if a majority of the coding can be handled by the end users, they have to have an estimate for how much it's going to cost. Some of this may simply be new LCNC technology, such as ongoing subscriptions, or a new database server. But the amount of work involved in coding, testing, installing, and training must be estimated. On top of this, management should keep maintenance in mind. A program will need to be supported (to correct bugs) and enhanced (to add or change features). That said, the good news is that users themselves can be empowered to handle a number of these tasks.

At this juncture it may not be clear what the total costs will be, given a specific LCNC platform or direction. Even working up estimates when there is a large amount of uncertainty in how the software will be developed can be dangerous; it may dramatically under- or over-estimate the costs. This can have the net effect of approving the project well ahead of knowing if it makes business sense, or of canceling it if the price tag is unreasonably high.

So, many times the best way to proceed is with a tentative "OK." Take next steps to clarify what's available, who's involved, and cost estimates. The members of the hub can move cautiously forward, taking small amounts of time to look at possibilities.

Figuring it out

It is important that members structure how they approach next steps. Although lively discussions are always welcome, they can many times end without moving things forward. One way to approach discussions is with an agenda and a structure to discuss the topics, including taking action.

1. **Identify.** Start by identifying if the organization currently uses LCNC platforms. Some LCNC capabilities are available on Google, Microsoft, Apple, and from other major tech firms. Additionally, identify what other packages are available to be used. This should be arranged in a list for review by the members.

2. **Refine**. For each of the platforms, the members should refine a list to the ones that hold the most promise. The features of each LCNC package, along with its compatibility with other formats and systems, in use should be investigated.

3. **Brainstorm**. With the candidates in mind, the team should open up discussion on solutions that will work. Not just the LCNC solutions, but other approaches that might offer a solution. These are suggestions from the members which are noted and displayed. If it is a video meeting, these can be written by the facilitator on the screen.

4. **Select**. From the list, members reduce the best solutions to the top three and rank them.

5. **Plan**. The team puts together a plan going forward. How will each selection be evaluated? Can trials be activated for each one where prototypes can be built to demonstrate the fit?

6. **Build**. Once a plan is in place, the next steps are to move forward. If possible, prototypes should be built with each LCNC platform under consideration. End users are involved in the actual implementation since most LCNC requires little coding experience. The prototypes can focus on the features that are the most needed. The result is then evaluated by the members and future users outside of the design hub.

7. **Test**. The prototypes are further tested. Any issues or suggestions are gathered for the next round of builds.

Final thoughts

In order to move forward with a solution that includes LCNC platforms, management should be included, as they ultimately possess the power to approve or reject the project. Presentations from the hub allow management to also be a part of the process. In this way, the team incorporates creative brainstorming into their selection process, allows for multiple approaches, and organizes its approach. With the group working in a collaborative mode, their teamwork will promote consensus and give members a chance to voice their evaluation.

20. Prototyping is the Most Fun Ever

Prototyping is modeling how the final piece of software is going to look and work. It's an iterative process which may start with a drawing on the back of a napkin, and find its way into a slideshow, and finally into a working application. This process of constructing prototypes fine-tunes the steps needed to produce the final software solution. Depending on the development process, the models can in fact be built using LCNC software—it's the ideal way to build the application since less code of the prototype is actually thrown away.

Prototyping begins after the team reaches a consensus on the type of application that is needed, the basic workflow, and a short-list of the types of output; once the top solutions are selected, initial prototypes can begin to be planned. Typically, the prototypes are placed into a testing environment, and hub members are given access to it. Sample data can be included and basic output can be demonstrated.

Of course, it's not a final product. But since prototypes are usually written using LCNC software, they can be changed relatively easily and updated to reflect comments from the stakeholders. Just what changes in the prototype and how it is changed are matters for discussion by the team members. Many times there is agreement, such as on the types of data used in calculations or the way information is displayed graphically. Other times, consensus is not easily reached. But even here, the prototype can be implemented with multiple approaches. This can be reviewed and common ground reached more easily.

A plan of attack

Before embarking on developing prototypes, the design team should determine which functions will be built out first and on which LCNC platform. For prototypes, the platforms used may be entirely different from the ones on which the final product is built, though ideally they use the same LCNC platform. This provides a seamless transition from the working prototype to the first production product ready for testing. (See section below for steps to take in selecting the right LCNC.) The process of planning the prototyping schedule involves the following considerations:

1. **Goals**. The overall goals of the project should be explained to and understood by the design team. If the types of features fall outside of the mandate, the team should consult with the advance team for guidance.

2. **Application outline**. The overall set of features should be outlined. This can be a simple spreadsheet that defines the different areas that are covered by the application. For example, if the application is simply an outward-facing website, then the outline should cover the different types of pages and information to be displayed.

3. **Priorities**. As Barry Boehm advised, the functions of each prototype iteration should be prioritized by how complicated they are to implement.[51] Complexity in the development process translates into an increased level of risk. If the LCNC platform cannot handle a particular set of functions, it can affect the choice of LCNC platform and how the application is implemented.

4. **Interfaces**. As part of the features, the team should identify any interfaces that the application needs to produce.

5. **Phases**. For larger endeavors, it may be necessary to break the project into phases. For example, Phase 1 may be limited to migrating current data into a new database and producing functional reports.

6. **Tasks**. For each of the phases and features, the design team should break the prototyping into tasks. For example, certain members may be responsible for defining the fields to be placed on the screen, while others refine the reports.

7. **Assignments**. The design team should look for members whose expertise most closely

matches the anticipated tasks. For example, IT members may be the best to work with security features, while accounting staff may be better suited for the general ledger entries the application produces.

8. **Preliminary estimates**. It is always helpful to produce an estimate of the amount of work involved. In the first prototypes, this may be very rough. A screen may take between 1-3 weeks on the first estimate, but as the team understands the platform better, this may change dramatically to 1-3 days per screen.

9. **Advance team**. Needless to say, the advance team should be consulted in developing the plan. Since resources in the organization will continue to be used on the project, keeping them in the loop is a prerequisite for continuity of staff and allocation of resources.

10. **Stakeholders**. As the prototyping begins, plans should be made to present the prototypes to the wider stakeholder community. These updates may span multiple departments and locations, but the advance team can assist with identifying who in the organization should be include and the best ways to demonstrate the prototypes.

Choosing the right LCNC platform

With a list of features and tasks, the design team can then evaluate which LCNC platform would work best for the prototyping phase. The purpose of the prototypes is to demonstrate how the final system will look and function: the workflow can be shown as a series of screens; the reporting can be organized in rough output; there may be required interfaces to existing systems in the organization. For example, if the application produces accounting entries, the LCNC must be capable of producing these in the correct format. Some platforms have built-in interfaces to other systems: where social media is used and needs to be updated, there may be built-in interfaces to Facebook, Instagram, WeChat, or Twitter.

When looking at the underpinnings of the software development platform, it's best to see what's available on the market and what experience the organization has with these packages. Not all packages will make sense. To hone in on the capabilities, the following questions should be addressed:

- Is the application solely a website displaying information?

- Does the application automate a workflow? If so, are there transactions that are initiated directly by the customer?

- What level of security is needed?

- To what extent are mobile devices used in the application?

- How is data collected and stored?

- Are customers located globally? Are different locations of the organizations involved in the processing?

- To what extent does the data need to be analyzed?

These types of questions can help the team select the LCNC platform that works the best for prototyping, with an eye to the final platform for the application.

User experience and user interface

Designing the order, composition, color scheme, and use of visual cues make up the field of *user experience* (UX). Composing screens and the flow of user input is a combination of artistic and technical skills.

On one hand, you want to make sure that you display the necessary fields on the screen and allow for the input of information in an orderly and logical way; you also want the frequent use of these screens to be easy. Too many fields crowded on a screen can become irritants and discourage users from being in the system, or result in information being entered incorrectly.

1. **Use existing themes**. Have a basic layout in mind for the screens. Refer to themes that you run into online, e.g., Material Design, Flat Remix, Ant Design, Grommet, and others. They can present a useable approach to the user interface.

2. **Keep it simple**. In designing the forms and pages, keep their presentation simple.

3. **Group related items**. Group related information together on a screen. For example, items related to name and address should be kept with other user information.

4. **Mobile layouts**. Keep in mind that the application being developed will no doubt be accessed by mobile devices; designs need to consider the portrait and landscape layouts on these smaller screens.

5. **User terminology**. Although the designers may describe information in technical terms, the prompts and descriptions should always be in the end user's language.

6. **Be consistent**. Once an approach to the layout and fields has been decided, keep it consistent across all screens.

7. **Prevent errors**. Wherever possible, check users' inputs to make sure no errors creep in. Where calculations can be done by the application, incorporate theses into the design.

These guidelines are helpful in the initial stages of designing the screens. In many cases, members of the design team can select the types of layouts from existing websites or systems.

Menus

Menus and submenus help the user navigate the application. Menus are typically placed at the top or are expanded from the three bars ("hamburger") in the top left on a mobile device. What displays when a user clicks on a menu item is a submenu with more detailed destinations. Keeping these consistent across the application allows the user to move freely to another page at any time. (If something interrupts data input, the user can be asked whether to save their data to complete at a future time.)

Forms and pages

Screens can be organized loosely into forms (which are designed for user input) and pages (which display information). Each may contain images, buttons, links, and pop-ups. These are part of most websites these days and find their way into all forms of *user interfaces* (UI). Both forms and pages may be grouped by the type of information that is being displayed or collected. These groupings are visually organized by tabs or sections on the screen. Clicking on the tab or section changes the screen to show the groupings of information. For example, in applying for a loan the applicant enters their name and address along with the same for other borrowers on the same loan application.

Screen overlays

When a user is on a page, there may be more information to show or collect. Instead of placing this directly on the screen, many platforms allow for this information to be placed into a box in front of the main screen. When the box is optional or may just display information, it is an overlay. When it requires the user's input, it is a modal. Both should be used sparingly, but they provide alternate means of displaying and accepting information, without leaving the page.

Workflow

For many systems, there is a workflow that links the screens to the final output. In the process, the input may be verified against the organization's records and the nature of each field. It passes to a series of approvals typically before arriving at a final set of data. Along the way there may be exceptions or errors that need special processing; these are typically flagged and handled in a separate set of steps.

Database set-up

Once collected, information must be stored securely and organized in a way that it is easily analyzed and updated. The centerpiece of any large application is its central set of databases. These can be relational databases in structure, where each field can be indexed and searched as needed (with the exception of the larger binary fields). Alternatively, the database may be a NoSQL type such as a document-oriented database. These databases specify the fields that are used to index the information, and store all information in a bundle (or document) that allows easy additions. The organization may already have a database it uses to warehouse information for reporting.

Many LCNC have built-in databases. By designing a form, the platform defines the fields, properties, and relationships, then places them into the database's schema. As users enter or edit information on the form, the LCNC platform updates the database accordingly. For example, Microsoft Access allows users to define their queries and also to pull information from the databases.

Reports

For many businesses and organizations, reporting is an important element of the automation process; reports may need to be generated on a daily, weekly, monthly, or quarterly basis, or available to staff in real time. As part of the prototyping, design team members may assemble the first cuts of reports that the final system will produce. The format and output will vary based on the needs of the organization.

Jumping in

With a plan in hand, prototyping can begin. These can be assembled and passed to members of the design team for comment and refinement. Although the actual approach can vary based on the members, some recommended steps in this process are:

1. **Sketch it**. Regardless of a person's artistic abilities, sketch out what the screens will look like, first in pencil or pen. Move fields around, group them together, and present it on a series of pages for the design team to review. This can be done using digital sketchpads, drawing programs, or photos of pen-and-paper drawings.

2. **Build it**. With the layouts in place, assemble the mock-ups using the LCNC platform. This can be as easy as dragging fields onto a screen, labeling them, defining their type and options, and saving them.

3. **Test it**. With the first set of screens, test them out. How easy are they to use? Are there fields that are missing?

4. **Refine it**. Try to reduce the number of fields and the amount of information that appears on the screens. Make sure all information needed for reporting is available.

5. **Repeat**. Once the prototype is complete, go back to step 2 and tweak the prototype.

With a prototype ready, it needs to be presented, tested, and refined by the design team.

Keeping the stakeholders in the loop

When prototypes are ready, bring in the advance team and stakeholders so they can begin to see how the application is shaping up. This can be done as a video or a set of slides showing the prototype. It is important to make sure that a wider audience is brought in early in the process. The prototypes may change, but the wider audience can be updated as these changes are implemented.

Final prototype

As the different functional prototypes are built, tested, and reviewed, they begin to form the full prototype of the system. Each functional element can be moved into a final prototype which goes through the same cycle of testing and review. This final prototype uses the LCNC platform chosen by the design team in consultation with the advance team. After the advance team and stakeholders have had a chance to view and comment on this prototype, it is ready to begin moving it into test environment for assorted end-users to test.

Final thoughts

The creation of a prototype is the next creative step in the iterative process of design. Hub members consider existing platforms as well as the needs of end users in building out the chosen LCNC platform. Both user interface and user experience must be considered for the final product or service to be successful. Several rounds of prototypes may be constructed, reconstructed pending feedback, and built back up before getting to the next phase of testing.

Part VII: Going Live

21. Crossing Over

As the prototypes are assembled and tested, they move from being test versions to being pre-production versions. They may still reside in the development environment, but the process of distilling their features also begins to simplify the underlying code. As it is introduced into the organization for their review and testing, these prototypes must also incorporate ways to educate users on how best to interact with the new system.

To this end, there are additions that can accelerate the process.

Video demonstrations

Putting together a series of demonstration videos that walk through the various parts (and potential quirks) of the system can be very helpful. It shows how the system is designed, the order that the steps take, how other staff members interact with the system, and final products such as reports and documents. These videos can be displayed on mobile devices or desktops. As these are prototypes in the final stages, the screen may change over time; the videos should be assembled simply.

Help screens

Designers may know every nook and cranny of the software by the time it begins to be rolled out, but first-time users will need help screens at some point, and including them will reduce calls to the support team. These may appear on the landing page of the system or as overlays when the user needs more information. On desktops, an overlay can appear when the cursor is over a field or menu item. On mobile devices, there may be an icon (e.g., a simple widget) that the user touches. If the design is intuitive, there is less need for lengthy help screens.

A separate "Getting Started" write-up is also helpful. It can be reviewed prior to getting on the new system.

Support

A support team should be assembled to assist with the initial rollout. These can be members of the departments where the system is being reviewed. Questions or bugs can be accompanied by screenshots and routed to the support team. Ideally there should be a Service Request database set up that allows users to enter their issues.

Warnings and errors

It is always best to enter spurious information to see how the software reacts. For example, can you enter a date earlier in the year or decimal points in the integer fields? Most LCNC platforms take care of this, but their warnings may be a bit too general. For this reason, the design team should make sure that warnings and errors are used as early in the process as possible.

Documentation

At a certain point, the team will need to consider general documentation for use by new hires and existing staff. It should be used to give an overview of what the software does, its requirements in terms of devices and operating systems, major functions, and examples of its use. This documentation is meant to give users an easy on-ramp to use the system.

FAQs

As part of the documentation, a simple Frequently Asked Questions (FAQs) sheet can inform users of many of the details and exceptions that are handled by the software.

Bringing in users

The design team can reach out to users to show them previews of what's in store. This can be done on a one-by-one basis, or in informal meetings with demonstrations. These previews are a good exposure to the types of questions users will have. Instead of waiting for these questions to appear at rollout, getting user feedback early can smooth out some of the rough spots in the interface and workflow.

Final thoughts

Testing of prototypes and pre-production versions of the final product or service makes sure the bells and whistles work for the most people, while also being streamlined and hopefully intuitive. FAQs and supporting documents can lessen the learning curve, while also lowering the burden of calls to support and service teams.

22. Rollout

Once the design teams have finalized their production code, it is time to roll out the software into phases for production. Keep in mind that not all solutions can be divided into phases; often it is better to implement the entire solution at the same time. For smaller applications, there should be no problem in deciding just how much to roll out. But even at this level, it is best to walk through the plans with the users for their input.

Rollout teams

Even though the hubs reflect the end users, organizations will no doubt have users outside of the design teams. These users remain largely outside the loop on the details of the features, functions, and planned implementation of the applications. Further, management and the formal organization should be brought into the timing of the rollout, to avoid times which may interfere with deliveries, critical services, holidays, and the like.

For this reason, a separate rollout team should be identified, one that is inclusive of management and formal teams, and takes the customers' needs into consideration. The rollout team is based largely within the formal organization, since resources, training, beta testing, and bug fixes must be considered.

Preview

In a series of presentations, the rollout team has the opportunity to preview the applications and gather recommendations from the design team for the phases that make the most sense. Since the development has been isolated to the development environment, the previews will largely include test data.

Where implementation can be isolated to a smaller subset of users, these users can be trained and can undertake additional beta testing in the development environment. Here transactions and data can be entered using existing production information (e.g., real customers and their transactions) in the development environment. Any mistakes made do not have an impact on the work of the organization. If a phased-in approach is used, then only the next phase can be tested in this manner.

Schedules

Based on the preview and feedback from end users, the rollout team can assemble their plans, including the timeframes, activities, evaluation periods, user teams, and review processes. These plans must be coordinated with the advance team. The design team should also be involved as there may be additional changes that are required. With LCNC the response time needed to change screens, reports, documents, workflow, and access is reduced, as there is less formal code to change. That said, those responsible for keeping the applications up-to-date should be aware of schedules.

Production environment

Once the schedules are accepted, the application can move forward. One of the first stops is moving it into the production environment, in phases if this is what is required. IT can move databases, set up user privileges, include the application into the operating system's authorized software, and grant access to folders on the main servers as needed. Here again, the rollout team can jump in to make sure the initial rollout in production looks good.

Configuration and training

Rollout needs to coordinate training of users long before the actual production dates. As intuitive as the user interface appears, users across the effected organization need to be brought into the fold and learn more about the system and its plans for implementation. For this, a formal training course can be constructed. These can be simple slides with screenshots, basic workflow, and examples. In addition to the presentations, it is useful to assemble instruction manuals both embedded in the application (where possible) and in a separate online document.

Where the application can be configured (for example, where users are able to preload templates for specific activities or customize their own screens, reports, and documents), the rollout team should work with users to have these in place. Interactive instruction works the best. IT can coordinate the configurations and training where additional resources are needed.

Train the trainer

For larger organizations, training the trainer works well. This may be one or more users who are given more detailed training with the aim of having them available in the organization for questions, to help resolve problems, and to become experts on the use of the applications. Some IT departments are already staffed to provide trainers.

Beta beta beta

It never hurts to conduct more beta testing with the trainers and members of the user community. There are always exceptions to the workflow, or new documents and reports that are needed. As part of their training, users can walk through more testing and how to report bugs or necessary changes to the design team.

Beta testing puts the system in the hands of end users to engage in production workflow with real data and situations. Results from the beta tests are compared to uncover any errors or to tweak the processing. Beta testing can run for a number of weeks, depending on staff availability. With the LCNC platforms, any bugs or enhancements that are identified can be easily updated in the system. Of course, care should be taken not to introduce bugs in the process, so usually this is done on a cycle.

Going parallel

Running both the new system alongside existing systems or procedures is a final test for accuracy and the handling of exceptions. Where there are existing systems, the results of the older system can be compared. Where manual processes are being replaced or updated, the rollout teams can manually direct users to view the new reports and documents against the manual results. It should be noted that running two systems in parallel takes additional time and resources, which is all the more reason to coordinate the rollout with management and the line supervisors when allocating the time to facilitate going parallel.

Can you roll out without going parallel? If the rollout team plans to skip the parallel step, then output from the application needs extensive verification as a final step during its initial days and weeks of going into production. For example, if the application produces end-of-day reports, these need line-by-line validation by the users. Or, if the software produces accounting entries for another system, these entries need to be first validated before they are sent up to the accounting system. These are extra manual steps that can slow down the process. However, there may be time savings that make it more feasible, or it may be too difficult to keep multiple production systems running in parallel.

Going live

Once the results of the parallel testing are available and show no problems, the application can be scheduled go live. Trainers can assemble the groups to view instructional slides and have interactive training where possible. Trainers can field questions and help during the ramp-up time. Once users are on the system, the trainers should be freed up to take questions, review problems, walk through the steps with individuals, and document any issues. Where users are dispersed, their workstations should be equipped with remote desktop connections to allow trainers a chance to see the work firsthand.

Bugs

There are always bugs, even with the best of testing. In these cases, the users should capture their work either through screenshots, saving results where possible, or jotting down notes on the steps that they took before the bug appears. Sometimes bugs are not bugs, but rather unique situations that require changes.

Trainers can be brought into this process to make sure that the user is entering information correctly or that they are accessing the correct data. Where members of the design team are available, they can view the screens and answer questions. IT can be brought in if the issues involve security, access, or capability with the operating system.

Ongoing changes

LCNC platforms are fairly easy to change even after going live. The design team can field suggestions and review exceptions to determine the priority and approach to making these changes.

Final thoughts

Rollout is pivotal to the software's implementation. It can be done in phases, or in parallel with existing systems. The rollout team should work with the design team to gather as much feedback as possible, as well as fix any potential bugs or address user errors. Implementation involves training users (and the involvement of the IT team) to deal with any technical issues that arise.

23. Maintenance: It's Never Over

Design teams can feel a sense of relief once their projects go live. Users are thrilled; management has another success; and there are pats on the back all around. But more often than not, beginning that first day in production, users will run into bugs and need changes. So, teams should be prepared to be available. Even simple questions that need answering get expedited to the team.

In a way, maintenance is its own project. It involves fixing the severe bugs quickly, restoring data from backups, or manually correcting incorrect calculations. It can also mean handling exceptions that weren't identified in the design.

The maintenance of typical software far exceeds the cost of developing it. Basic maintenance consumes about 40% to 80% (60% average) of software costs. Of these, enhancements (which add functionality) are responsible for about 60% of costs, with bug fixes on most systems comprising about 17%.[52] LCNC significantly reduces these costs since the actual coding is simplified. That said, you should be prepared for continued updating of the software.

Support team

From the design team, a smaller team should be identified to handle the issues that arise. They may be simple questions or involve going into the code and fixing bugs. These teams can be part of the formal organization and taken on by IT. Most of the time, users will ask their peers when they run into problems. These peers can be the same trainers that rolled out the software, or they may be the ones that felt the most comfortable using it. In either case, the support team is well advised to seek out these coworkers and ask if they need more information or training; they can be the first line of defense.

Service requests

With the software in use, it's always best to allow users to create service requests when they run into a problem, have an idea, or are stuck for whatever reason. These requests can be embedded into the software or run separately on a separate application. The goal is to walk the user through the information that the support team will need.

Trainers as experts

With trainers already up to speed on the software, they can be used to help users with questions and problems they may encounter.

Prioritizing

As requests come in for fixes or changes, the support staff needs to decide on their priority. Obviously crashes, serious errors, and incorrect calculations need to be attended too quickly. With LCNC it is easier to spot the source of these problems. The support staff are not always the ones that wrote the code, but the coders themselves can be available to answer questions or guide the staff.

Once the problem area has been identified, it is important to make the changes first in the current version in the development environment. Tests should be run by the support staff to make sure that the fix doesn't cause other problems. Larger fixes require more extensive testing. Instead of rushing to get the problem solved, the support team should make sure that the fix works and does not cause other issues. This can then be rolled out.

On-the-spot fixes

It's not always possible to fix issues in the development environment and bring it over into the production environment. In cases where there is an urgency to rectify the problem, fixes are made in production first, tested by the support team and the user, and then made more permanent. This is always a judgment call, but the situations will dictate the urgency of the fix.

Collecting bugs

For those problems that are minor, the staff can collect bugs for a future release. These bugs fall into the category of bothersome or inconvenient, but not serious. For example, prompts on the screen may have a misspelling, or formats are not consistent. The support team collects these and their service requests for future releases.

Enhancements

Maintenance is all about enhancements. It's only when a system is used that the staff realize that new features and functions would have a great boost to performance or added value to the business. In general, enhancements can be prioritized much like fixes, with the important ones getting more immediate attention and the ones that are "nice to have" postponed for a later release.

Releases

Releases are new versions of the software with added functionality, corrected bugs, and new steps. In most cases, a brief synopsis of what is included in the release is all that it required. But deciding on what goes into each release should be a joint decision of the support team and the users.

Final thoughts

There is never a final release of a new product or software. As users begin actually using the system there will always be things in need of an update. Users may find bugs, better solutions to streamline workflow, or give request enhancements to the basic functionality. Ongoing support teams and the IT department are always needed for maintenance.

Part VIII: Quick Guide

24. Summary

Software development has changed dramatically for enterprises over the last several years. What used to be a straightforward process of writing up requirement specifications that could be implemented by software engineers now involves a quicker, more collaborative process. In addition, the layered, full-stack approach of building an application from the ground up has been replaced by low-code/no-code development platforms with pre-built components.

Not all new development lends itself to this new approach to digital transformation, but for those that do, the process can accelerate the implementation and produce new systems that are better attuned to the needs of its users. As these needs change, the LCNC platforms can accommodate enhancements more easily.

The process of incorporating LCNC solutions into the enterprise can be further improved by organizing the informal organization into design hubs. These hubs, with the guidance of management advance teams, are a network of individuals across departments and boundaries that are engaged to come up with creative solutions, and are run by those with the most knowledge of the work at hand. Since products and services these days are more complex and must be customized for different locales, the design hub has greater flexibility, can explore the issues that are being targeted, and can involve users and customers from the beginning.

The design hubs are further empowered using digital hubs and a structured approach to design.

Digital hubs establish a communications platform that incorporates real-time messaging with asynchronous forums. Instead of needing to be face-to-face in conference rooms or off-site, digital hubs allows members from different locations to be involved. Their input is captured in a central set of folders or channels.

In addition, by following a structured set of collaborative design steps, the design hubs can examine the issues, refine their presenting problems, and measure the impact before adopting a solution. The solutions themselves are structured, starting first with brainstorming activities, selecting the best candidates, then interacting through a series of prototypes to refine the software solution. The prototyping includes planning, building, and testing phases. Once the prototypes show promise, they can be transitioned to the final software. Along the way, end users can be involved in the design and coding. Since LCNC platforms do not require low-level languages, there are opportunities for users to learn the LCNC environment and to experiment with different layouts, documents, views, and reports.

Quick guide

Whether the goal is to develop new software, to add major enhancements to existing systems, or to simply understand the issues more clearly, the basic steps of low-code/no-code design can be summarized in the below nine steps:

1. ***Build an effective bridge between the organization's hierarchy and informal workgroups.***

The management of an organization needs to form an advance team when there are major issues or goals that require expanding its digital footprint. This team is formed from different departments and levels in the organization to reflect key stakeholders. The advance team clarifies the direction that is needed and guides the formation of the central design hub to lead the project. The mandate should be clear across the departments. One of their first tasks is to recommend a group of people who in turn build a main design hub, which will handle the design and development process.

2. ***Engage a network of individuals to form the design hub which will collaborate on the design and coding of the software application.***

With the advice of the advance team, the design hub forms with the inclusion of IT, end users, stakeholders, and potentially customers with the purpose of delivering solutions as defined through the advance team. Along the way, this hub may encounter other issues and goals, but their work begins with the mandate they are given. The core group introduces the collaborative design process that will be used, along with suggestions for rules, venues, and approaches.

3. ***Organize the digital design hub through the introduction and use of digital communications platforms.***

With the design hub in place, technology must allow for real-time notifications, online forums, shared documents, digital conversations, video presentations, video conferencing, and priority e-mail. In addition to face-to-face meetings, the digital nature of the communications allows members to communicate remotely, from different locations, and at different times to promote continuity and exchange of ideas.

4. ***Structure work using the collaborative design process to gain a fuller understanding of the issues, to promote more creative solutions, and to include stakeholders early in the design process.***

By structuring the design and coding with the seven-step process of collaborative design, designers are better able to define the targets, understand how they impact the organization, measure the occurrences and severity where possible, and refine a final set of targets that will be the centerpiece of the final software. Informal discussions are encouraged but the end point should be action plans that lay out the next steps. With a prioritized list of targets, the design hub brainstorms solutions. These are refined and the top solutions are selected for the next steps.

5. ***Evaluate and select the low-code/no-code development platform that best fits the requirements of the organization.***

Selecting the best LCNC platform for the organization is based on its expense, technical fit, ease of implementation, security, and support. Since this platform can be used on other projects or to expand the digital footprint of the organization, the design hub should include other stakeholders in its selection process.

6. ***Use low-code/no-code platforms to prototype the solutions with key stakeholders.***

The design hub then creates a series of prototypes, each focused on a different aspect of the overall solution. Once the prototypes are validated by the end users and stakeholders, the LCNC application can be assembled through iterative steps. Results of each cycle of testing should be submitted to the advance team for their review and guidance.

7. *Transition to the production version of the software with the inclusion of training, documentation, configurations, and schedules.*

With the successful prototypes, the production version of the code can begin when major issues have been resolved and the design hub receives approval from the advance team and management. The production version can be rolled out in phases with alpha and beta testing environments that incorporate current inputs and data. Each phase is reviewed and tested by the end users for completeness, ease of use, and compliance.

8. *Go live!*

The final software application is then moved into the production environment after training, documentation, and management briefings have been completed.

9. *Prepare for ongoing maintenance that includes customizations, user assistance, and new versions.*

As part of the rollout, support staff should be included in the final testing phases, training, and documentation. The design hub should remain in place until the support staff is able to fully handle requests for changes or corrections.

25. Endnotes

1   "What You Need To Know About The Low-Code Market," *Forbes*, 02/17/2019.

2   "Global And United States Low Code Development Platform Market Size, Status And Forecast 2020-2026," *360 Research Reports*, 08/20/20. https://www.360researchreports.com/global-and-united-states-low-code-development-platform-market-16173276

3   Dale E. Zand, "Collateral Organization: A New Change Strategy," *The Journal of Applied Behavioral Science*, vol. 10, no. 1, pp. 63- 89. M. E. Shaw, G. H. Rothchild, and J. F. Strickland, "Decision process in communication nets," *Journal of Abnormal and Social Psychology*, 1957, no. 54, pp 323 - 330. A. Bavelas, "Communications patterns in task-oriented groups," *Journal of Acoustical Society of America*, 1950, no. 22, pp 725 - 730.

4   Alex Pentland , *Social Physics: How Social Networks Can Make Us Smarter.* Penguin Books, 2014.

5   Quentin Hardy, "The New Workplace Is Agile, and Nonstop. Can You Keep Up?" *The New York Times*, 11/25/16.

6   M. Mitchell Waldrop, *Complexity*, Simon & Schuster, New York,1992, p.16.

7   Nancy Rytina, "Estimates of the Legal Permanent Resident Population in 2011," *Population Estimates July 2012*, Department of Homeland Security. https://www.dhs.gov/xlibrary/assets/statistics/publications/ois_lpr_pe_2011.pdf. Mitra Toossi, "A Look At The Future Of The U.S. Labor Force To 2060," *U.S. Bureau of Labor Statistics*, September 2016. https://www.bls.gov/spotlight/2016/a-look-at-the-future-of-the-us-labor-force-to-2060/home.htm

8   Mitra Toossi , "A new look at long-term labor force projections to 2050," *Office of Occupational Statistics and Employment Projections*, Bureau of Labor Statistics, September 2006.

9   Bryan Bassett, *U.S. Mobile Worker Population Forecast, 2020–2024*, IDC publication #US46774020, Aug 2020.

10   *Projections overview and highlights, 2019–29*, Bureau of Labor Statistics, September 2020. https://www.bls.gov/mlr/2020/article/projections-overview-and-highlights-2019-29.htm

11   *Mobile Workers Will Be 60% of the Total U.S. Workforce by 2024, According to IDC*, IDC publication # US46809920, September 2020

12   C. Ford and D. Ogilvie, "The role of creative action in organizational learning and change," *Journal of Organizational Change Management*, 1996, 9(1), pp 54 - 62.

13   Waldrop, *Ibid.*, pp 144- 147.

14   G. Doherty, N. Karamanis and S. Luz, "Collaboration in Translation: The Impact of Increased Reach on Cross-organisational Work," *Computer Supported Cooperative Work:* The Journal of Collaborative Computing (JCSCW),21(6) 2012, 525–554.

15   "Boeing Hits a Milestone," *The Wall Street Journal*, 6/8/12.

16   John P. Kotter, *Accelerate: Building Strategic Agility For A Faster-Moving World*. Harvard Business Review Press. Boston, MA. 2014.

17   Jay R. Galbraith, *Organization Design,* Addison-Wesley Publishing, 1977, p. 112 - 114.

18   *Ibid,* p. 140 - 143.

19   "Why Microsoft's Nadella may succeed where Ballmer failed," CNBC, 3/11/14. Also, "Microsoft's New CEO: This Is The Big Culture Change We Need to Change," *Business Insider*, 2/20/14.

20   https://www.statista.com/chart/16903/microsoft-stock-price-under-satya-nadella/

21   Christian Reuter, *Emergent Collaboration Infrastructures: Technology Design for Inter-Organizational Crisis Management*, Springer, 2014.

22   Kotter, *Ibid.* p.94 - 97.

23   *The New York Times Innovation Report*, 2014, p. 31. https://www.scribd.com/doc/224332847/NYT-Innovation-Report-2014.

24   Lucas Alpert, "The New York Times Lays Out Plans to Restructure Newsroom," Wall Street Journal, 1/17/17.

25   K. McKenna, A. Green, and M. Gleason, "Relationship formation on the Internet: What's the big attraction?" *Journal of Social Issues* 58, 1 (2002)

26   Burr Settles and Steven P. Dow, "Let's Get Together: The Formation and Success of Online Creative Collaborations," *ACM CHI Conference*, (2013).

27   M. Burke and B. Settles, "Plugged in to the Community: Social Motivators in Online Goal-Setting Groups," *International Conference on Communities & Technologies (C&T)*, pp 1-10. ACM (2011).

28   Edgar Schein, *Process Consultation: Its Role in Organization Development*. Addison-Wesley Publishing, Reading, Mass., 1969, pp. 46 - 48.

29   Don Norman, *The Invisible Computer,* MIT Press, Cambridge Mass., 1998, p. 185.

30   https://designthinking.ideo.com/

31   H. Lieberman, F. Patermp, and V. Wulf (Eds.), *End-User Development*. Kluwer/ Springer, 2006.

32   Barry Boehm, "A Spiral Model of Software Development and Enhancement," *ACM SIGSOFT Software Engineering Notes*, August 1986.

33   Richard Beckhard, *Organization Development*. Addison-Wesley Publishing, Reading, Mass., 1969. Warren Bennis, *Organization Development: Its Nature, Origins, and Prospects*. Addison-Wesley Publishing, Reading, Mass., 1969.

34   Newton Marguiles and Anthony Raia, *Conceptual Foundations of Organizational Development*, McGraw-Hill, New York, 1978.

35   W. L. French, "Organization Development: Objectives, Assumptions and Strategies," *California Management Review*, 12(2):23-24, 1969.

36   Harrison Roger, "When Power Conflicts Trigger Team Spirit," European Business, Spring (1972), pp 57 - 65.

37   https://www.inc.com/michael-schneider/google-thought-they-knew-how-to-create-the-perfect.html

38   http://www.nytimes.com/2016/02/28/magazine/what-google-learned-from-its-quest-to-build-the-perfect-team.html

39   Amy J. Ko, et. al., "The State of the Art in End-User Software Engineering," *ACM Surveys Vol. 43, No. 3*, Article 21, April 2011.

40   Kotter, *Ibid.* pp 94 - 97.

41   S. Mirri, M. Roccetti and P. Salomoni, "Collaborative design of software applications: the role of users," *Human-centric Computing and Information Sciences*. Volume 8, No. 6 (2018).

42   A. Ebersbach, M. Glaser, and R. Heigl, *Social Web*. Konstanz, Germany, UVK, 2008.

43   H. Hippner, "The Importance, Application and Potential of Social Software," In K. Hildebrand & J. Hofmann (Eds.), *Social Software*(pp. 6–16). Heidelberg, Germany, (2006).

44   O. Hazzan and Y. Dubinsky, *Agile Anywhere*, SpringerBriefs in Computer Science(2014).

45   M-C Boudreau, K. D. Loch, D. Robey and D. Straud, "Going global: Using information technology to advance the competitiveness of the virtual transnational organization," *Academy of Management Executive* 12(4) 1998, pp 120-128.

46   K. Schmidt and L. Bannon, "Taking CSCW Seriously: Supporting Articulation Work," *Cooperative Work and Coordinative Practices*, 1(1) (1992)., pp 1–33.

47   G. Doherty, , N. Karamanis, and S. Luz, "Collaboration in Translation: The Impact of Increased Reach on Cross-organisational Work," *Computer Supported Cooperative Work:* The Journal of Collaborative Computing (JCSCW), 21(6) (2012), pp 525–554.

48   Kotter, *Ibid*. Based on John Kotter's concept of the types of teams that can influence change in organizations.

49   Jack Fordyce and Raymond Weil, *Managing with People,* Addison-Wesley Publishing Co., Reading, MA, 1971, pp 137 56.

50   Lewin, *Ibid., pp. 34- 46.*

51   Boehm, *Ibid.*

52   Robert Glass, "Frequently Forgotten Fundamental Facts about Software Engineering," IEEE Software, Volume: 18, Issue: 3, May 2001.

www.ingramcontent.com/pod-product-compliance
Lightning Source LLC
Chambersburg PA
CBHW052341210326
41597CB00037B/6215